High Priorities: Conserving Mountain Ecosystems and Cultures

DEREK DENNISTON

Ed Ayres, *Editor*

WORLDWATCH PAPER 123
February 1995

THE WORLDWATCH INSTITUTE is an independent, nonprofit environmental research organization based in Washington, D.C. Its mission is to foster a sustainable society—in which human needs are met in ways that do not threaten the health of the natural environment or future generations. To this end, the Institute conducts interdisciplinary research on emerging global issues, the results of which are published and disseminated to decisionmakers and the media.

FINANCIAL SUPPORT is provided by the Geraldine R. Dodge Foundation, W. Alton Jones Foundation, John D. and Catherine T. MacArthur Foundation, Andrew W. Mellon Foundation, Edward John Noble Foundation, Pew Charitable Trusts, Lynn R. and Karl E. Prickett Fund, Rockefeller Brothers Fund, Surdna Foundation, Turner Foundation, U.N. Population Fund, Wallace Genetic Foundation, and Frank Weeden Foundation.

PUBLICATIONS of the Institute include the annual *State of the World*, which is now published in 27 languages; *Vital Signs*, an annual compendium of the global trends—environmental, economic, and social—that are shaping our future; the *Environmental Alert* book series; and *World Watch* magazine, as well as the *Worldwatch Papers*. For more information on Worldwatch publications, write: Worldwatch Institute, 1776 Massachusetts Ave., N.W., Washington, DC 20036; or FAX (202) 296-7365.

THE WORLDWATCH PAPERS provide in-depth, quantitative and qualitative analysis of the major issues affecting prospects for a sustainable society. The Papers are authored by members of the Worldwatch Institute research staff and reviewed by experts in the field. Published in five languages, they have been used as a concise and authoritative reference by governments, nongovernmental organizations, and educational institutions worldwide. For a partial list of available Papers, see page 81.

DATA from all graphs and tables contained in this book, as well as from those in all other Worldwatch publications of the past year, are available on diskette for use with Macintosh or IBM-compatible computers. This includes data from the *State of the World* series, *Vital Signs* series, Worldwatch Papers, *World Watch* magazine, and the *Environmental Alert* series. The data are formatted for use with spreadsheet software compatible with Lotus 1-2-3, including Quattro Pro, Excel, SuperCalc, and many others. Both 3 1/2" and 5 1/4" diskettes are supplied. To order, send check or money order for $89, or credit card number and expiration date (Visa and MasterCard only), to Worldwatch Institute, 1776 Massachusetts Ave., NW, Washington, DC 20036. Tel: 202-452-1999; Fax: 202-296-7365; Internet: wwpub@igc.apc.org.

Printed on 100-percent non-chlorine bleached, partially recycled paper.

Table of Contents

Tables and Figures

ACKNOWLEDGMENTS: Numerous people made this report possible. For their professional critiques and insights, I am deeply grateful to Jayanta Bandyopadhyay, Alton Byers, Gabriel Campbell, Bob Davis, James Enote, Gary Fleener, Larry Hamilton, Jack Ives, Rodney Jackson, Nels Johnson, Juan Mayr Maldonado, Bunny McBride, Manjari Mehta, Bruno Messerli, Francis Ojany, Egbert Pelinck, Jane Pratt, Martin Price, John Ryan, Jim Underwood, Peter Weber, Carrie Wood, and Jim Wood. I am especially indebted to Elizabeth Byers for reviewing two near-final drafts on short notice. I thank everyone at The Mountain Institute for their enthusiastic help. For infusing me with renewed inspiration to write about mountain cultures and ecosystems, I am grateful to the other participants in the NGO Workshop on the Mountain Agenda, held at Spruce Knob, West Virginia. And I would have never completed this project without the clear vision and continuous love of my best friend and wife, Chris Denniston.

Many colleagues at Worldwatch Institute merit special recognition: Susan Chandler for volunteering to help with the early research; Anne Platt for her efficient assistance; Anjali Acharya for helping me tie up loose ends; Sandra Postel for offering steady oversight throughout the project; Nick Lenssen for meticulously proofing the endnotes; Chris Bright for proofing the entire manuscript; Denise Byers Thomma for guiding the manuscript through the production process; and Ed Ayres for becoming a full partner with me in editing and polishing the paper.

Finally, I would like to acknowledge the women and men throughout the world's mountain communities doing conservation and sustainable development work. Although I was able to cite only a small number of these projects, their efforts are the inspiration for this report.

DEREK DENNISTON is Research Associate at the Worldwatch Institute, and coauthor of the Institute's *State of the World 1995* report. His research focuses on mountain environments and conserving biological diversity. He is a graduate of Principia College, where he studied world history.

One-Fifth of the World's Landscape

Mountains make up one-fifth of the world's landscape and are home to at least one-tenth of the world's people. An additional 2 billion people depend on mountains for much of their food, hydroelectricity, timber, and mineral resources. All told, fully half of the world's people, as well as a surprisingly large share of its biological diversity, depend on mountain watersheds for fresh water. In an era of increasing water scarcity, perhaps no category of the earth's major biomes has greater value for geopolitical—not to mention environmental—security. Yet, in the deliberations of governments and organizations worldwide, the fate of the mountains has been largely ignored.[1]

For millennia, mountains have evoked powerful emotions— fear, awe, and reverence—among those who lived within sight of them. To indigenous peoples on every continent, mountains were the places where gods dwelled. To romantic writers and artists of the unfolding modern era, they were universal symbols of human aspiration. Even today, more than one billion people consider a mountain sacred. But in an increasingly industrialized and urbanized world, that leaves a growing majority of humanity whose mental disconnection from the sources of their sustenance has made them indifferent to these sources. Relatively few have any real awareness of the critical natural services mountains provide; even fewer are aware that mountains harbor an exceptionally rich and complex life of their own. Unlike the oceans or tropical rain forests, mountains have never had their own scientific discipline, or even a movement to broadcast the grave threats facing them and their peoples, such as Jacques Cousteau has fostered for the oceans.

Except for those whose interest is recreational or religious, mountains have remained on our mental margins.[2]

Whether in cloud forests or alpine grasslands, on windswept promontories or along glacier-fed streams, what mountain ecosystems have in common—what makes them a distinct biome unto themselves—is the combined effect of rapid changes in altitude, climate, soil, and vegetation over very short distances. Biologically, their high diversity—including prolific concentrations of endemism, or species found nowhere else—leaves them vulnerable to losses of whole plant or animal communities. And culturally, the fact that most mountain peoples are ethnic minorities, outside the dominant cultures of the plains, leaves their regions poorly represented in the centers of political or commercial power where much of their fate is determined. In the surrounding lowlands, millennia of intensive human use have led to steadily increasing biological impoverishment and cultural homogenization. Mountain peoples, in their vertical archipelagos of human and natural variety, have become the guardians of irreplaceable global assets.

There is a dangerous disproportion, therefore, between the great importance of mountain ecosystems and cultures and the attention they receive in national governments and international organizations—a disparity that increases the risks that now confront both the mountains themselves and all those who rely on them. Moreover, those risks are magnified by the fact that mountain communities—human and natural—have unique vulnerabilities, stemming from the same characteristics that lead most people to circumvent them: their steepness, isolation, and marginality.

These vulnerabilities expose mountain communities to pervasive degradation. And upland losses can have immense downhill ramifications. The degradation is being caused primarily by three external threats from the expanding world economy. First, the prevailing patterns of inequitable land ownership and population growth in many mountainous developing countries have led to scarcities of arable land, and growing strain on subsistence communities. Second, these communities and their surrounding natural assets have become targets of intensive resource extraction. And finally, mountain attractions—from the ski slopes in Canada and Chile to trekking routes in Kenya and Nepal—have generated a global boom in tourism and recreation. All over the world, mountain peoples face increasing cultural assimilation, debilitating poverty, and political disempowerment.

Because of these threats, mountains have become critical proving grounds for sustainable development. Reversing these threats will be impossible without a deeper understanding of the profound interdependence between mountains' human and natural communities. And, while ultimate solutions for the mountains will demand changes as revolutionary as those required for reforming the global economy itself, there are shorter-term measures—at the community, regional, national, and global levels—that are both imperative and politically achievable.

The Vertical Dimension of Diversity

Mountains and high plateaus exist at every altitude and terrestrial latitude. (See Map.) They cover at least 30 million square kilometers and in 1995 were home to at least 570 million people. Of the world's current roster of 185 countries, only 46 have no mountains or high plateaus—and most of those are small island nations. More than one hundred mountain ranges in about 80 countries are large enough to have their own names. Although even basic data on the area and population of many ranges are hard to come by, what are available show that mountains contain wide geographic and demographic diversity. About half of the world's mountain peoples are concentrated in the Andes, the Hengduan-Himalaya-Hindu Kush system, and dispersed African mountains. (See Table 1.) In contrast with the sparse populations of Northern mountains, portions of some tropical ranges—such as parts of the highlands of Papua New Guinea, the Vale of Peshawar in northern Pakistan, the Virunga volcano region of Rwanda, and Mount Kenya—have more than 400 people per square kilometer.[3]

For a majority of people, mountains are not home but are

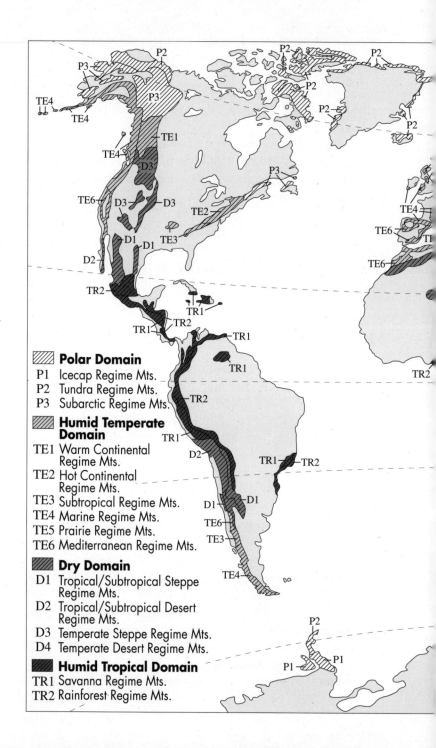

Polar Domain
P1 Icecap Regime Mts.
P2 Tundra Regime Mts.
P3 Subarctic Regime Mts.

Humid Temperate Domain
TE1 Warm Continental Regime Mts.
TE2 Hot Continental Regime Mts.
TE3 Subtropical Regime Mts.
TE4 Marine Regime Mts.
TE5 Prairie Regime Mts.
TE6 Mediterranean Regime Mts.

Dry Domain
D1 Tropical/Subtropical Steppe Regime Mts.
D2 Tropical/Subtropical Desert Regime Mts.
D3 Temperate Steppe Regime Mts.
D4 Temperate Desert Regime Mts.

Humid Tropical Domain
TR1 Savanna Regime Mts.
TR2 Rainforest Regime Mts.

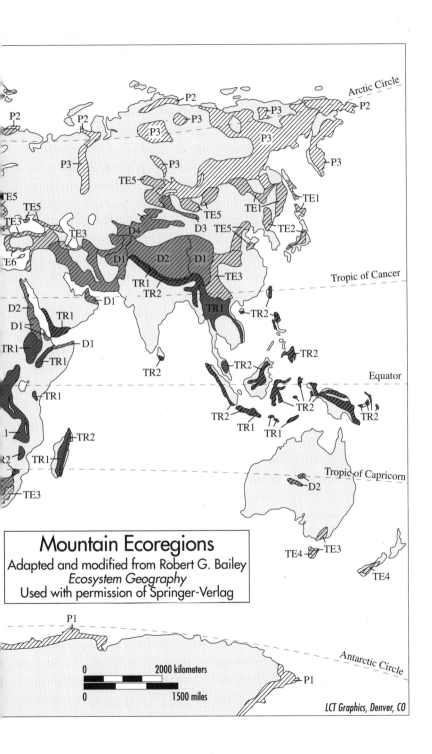

Mountain Ecoregions
Adapted and modified from Robert G. Bailey
Ecosystem Geography
Used with permission of Springer-Verlag

0   2000 kilometers

0   1500 miles

LCT Graphics, Denver, CO

TABLE 1
Area and Population, Selected Mountain Ranges and High Plateaus[1]

Region	Area (thousand square kilometers)	Population, Circa 1990 (millions)
N.E. Siberia and Russian Far East	3,813	1.7
Himalaya[2]	3,400	121
African Mountains and Highlands[3]	3,000	100
Tibetan Plateau[4]	2,500	10
Andes	2,000+	65
Western Canadian Ranges[5]	1,365	3.3
Antarctic Ranges	1,346	0
Alaskan Ranges[6]	1,060	0.17
Other Former Soviet Ranges[7]	836	2.4
U.S. Rockies[8]	818	3.4
Central Asia and Kazakhstan	522	11
Sierra, Calif. Coastal, and Cascades	380	17.1
Brazilian Atlantic Range[9]	300	25
Alps	240	11.2
Caucasus	179	7.7
Appalachians and Adirondacks	176	10.0

[1]Figures are not strictly comparable, due to varying definitions of mountains and plateaus. [2]Includes Hengduan, Hindu Kush, and Karakorum. [3]Includes East African mountains and highlands, Atlas mountains, isolated West African mountain peaks, dispersed southern African mountains, and mountains and highlands of Madagascar. [4]Includes only the portion within the People's Republic of China, containing the Tibetan Autonomous Region, and the following autonomous political units in these provinces: six prefectures in Qinghai, one prefecture and one county in Gansu, two prefectures in Sichuan, and one prefecture in Yunnan. [5]Includes portions of Alberta, British Columbia, Northwest Territories, and Yukon Territory. [6]Includes Ahklun Mountains, Alaska Range, Aleutians, Brooks Range, Pacific Coastal, Pacific Gulf, Seward Peninsula, Upper Yukon mountains, and Yukon Intermontane Plateaus. [7]Includes Altai, Sayan, Urals, high latitude mountains, Crimea, and the Ukranian Carpathians. [8]Includes New Mexico-Arizona and Nevada-Utah mountains. [9]Includes Chapada Diamantina, Serra do Espinhaço, Serra da Mantiqueira, Serra do Mar, and Serra Geral.

Source: Compiled by Worldwatch Institute from sources cited in endnote 3.

familiar features of the landscape—whether the Himalaya as seen from Kathmandu, the Andes Cordillera from Lima, or the Cascade Range from Seattle. Often the nearest range is not very far. Yet, common as they are, these features remain—for the world's increasingly urban population—at the margins of consciousness. Just as they so often do in art, mountains in the real world of commerce and politics are likely to serve as a kind of out-of-focus backdrop for more "important" subjects. And when mountains become the objects of direct attention at all, they are generally seen as unmoving and unmovable—as immutable. That such massive monoliths of rock and ice could also include highly complex, ever-changing, and *vulnerable* ecosystems—and be home to diverse cultures—seems inconceivable.

The first step toward reversing the degradation of mountain ecosystems and cultures, then, is to move them from the margins of public consciousness to a more central place on national and international agendas. Part of that task is to fill in the enormous gaps in what is known about the effects of human activities. These gaps exist, in part, because mountains have been marginalized by the artificial compartmentalization of knowledge into scientific disciplines and of economies into sectors. This segmentation of knowledge has tended to discourage any integrated understanding of complex mountain ecosystems, in which geological, meteorological, biological, cultural, and economic phenomena are too interdependent to be studied or measured in isolation. Not surprisingly, there has been a tendency—in government planning, nongovernmental activism, and even in scientific research—to carve interdependent ecosystems into artificial fragments. Often, critical linkages between upper and lower watersheds, mountain forests and grasslands, and upland and lowland communities have been ignored. This ignorance, for instance, has helped to perpetuate the popular myth that mountain ecosystems exist only above treeline. But the most basic factor keeping mountain issues from their rightful place on environment and development agendas may be a fundamental misunderstanding of the terrain. Mountains are not simply steeper or higher versions of the same ecosystems that occur in the flatlands, but are distinctive systems providing uniquely valuable resources—resources that are, because of our unsustainable exploitation of them, also threatened.

What makes mountains different from all of the planet's other biomes is their vertical dimension. Mountain landforms are largely a result of ancient and continuing collisions of tectonic plates, which account for the great instability of their geology—and the notorious frequency with which they produce such natural hazards as volcanic eruptions, mass erosion, landslides, glacial-lake outbursts, earthquakes, avalanches, debris flows, sedimentation, and floods. Far from the immutable landforms they appear to be from a distance, mountains are in continual upheaval—collapsing bridges, washing out roadbeds, and breaching dams. These physical instabilities not only add major costs to the building of human-made structures, but can make human settlement dangerous. On May 31, 1970, for example, an earthquake caused a large mass of ice to tumble from Mount Huascarán in the Peruvian Andes, creating a massive mudflow that hurtled toward the town of Yungay 12 kilometers away. Fifteen minutes later, 18,000 villagers were dead.[4]

The complexity of mountain topography—including variations in elevation, slope, and orientation to the sun—create large variations in temperature, radiation, wind, moisture availability, and soils over very short distances. This physical diversity leads to comparable variety in vegetation and animal life. In fact, one of the most defining ecological characteristics of mountains is that the rise in elevation is sufficient to produce "altitudinal zonation"—elevational belts (or zones) of climates, soils, and vegetation. One recent geographic analysis shows 23 percent of the earth's landscape with this characteristic. Many mountain ranges include high plateaus that lack this rise in elevation. However, high plateaus are usually included in discussions (including this one) of mountains because their lofty elevation induces cold, thin air and short growing seasons, and thus the low biomass productivity that is common to many steep-sloped environments. Mountains include both the wettest place on earth—Cherrapunji, India, which receives almost 12 meters of rain annually—and one of the driest—Chile's Atacama Desert, which has had no measurable precipitation in at least 27 years.[5]

An indication of the extreme climatic variation in mountains is that 100 meters of elevation gain is roughly comparable to 100 kilometers change in latitude. With high daily and seasonal variability in solar radiation, temperature, and precipitation, weather becomes a key factor in the distribution and variety of soils. Another key factor is the effect of gravity on steep inclines, which can produce massive movements of weathered rock and soil—stripping some kinds of terrain bare while accumulating rich deposits in others. These variations, in turn, extend the range of possible biological and human adaptations. At high altitudes, extreme temperature fluctuations induce native plants and animals to develop special survival mechanisms. Even the shade of a single rock can provide a unique microhabitat for an alpine wildflower found nowhere else. Similarly, a mountain farmer might use completely different seeds, cultivation, and harvesting practices for fields that differ by as little as 50 meters in altitude.[6]

The first step toward forestalling this degradation is to move mountains from the margins of public consciousness to a more central place on national and international agendas.

A key component of weather is the amount of moisture produced, and while this varies widely, mountains are generally among the wetter places on the planet, functioning as the earth's water towers by receiving much of its precipitation. By intercepting the global circulation of air, they force air upward, where it condenses into clouds that provide rain and snow. Thus, most of the world's major rivers form in mountains, giving the upper portions of watersheds immense value for environmental and geopolitical security. About 1 billion Chinese, Indians, and Bangladeshis, for instance, depend directly on water that originates in the Himalaya. At least 250 million people depend on water from the mountains of Africa. Most of California's 31 million residents depend on rivers originating in the Sierra Nevada. Altogether, about half of the world's human population relies— for drinking, domestic use, irrigation, hydropower, industry, and transport—on water captured in mountains.[7]

The flow of water through mountain watersheds is a timeless phenomenon that belies the real fragility of these environments—and their vulnerability to human disturbance. Since marked topographical relief produces unstable surfaces, mountain soils are usually young, shallow, and poorly anchored. Low temperatures cause slow soil formation and vegetation growth, while gravity-powered erosion accelerates silt and sediment movement. Because a doubling in water speed produces an eight- to sixteen-fold increase in the size of the particles that flowing water can transport, the erosive power of rapid runoff in mountains is immense and can create sudden and irreversible losses of soils. And compared with more productive lowland environments, mountain ecosystems are typically less able to recuperate from heavy losses of soil or vegetation. In high or cold-stressed mountain environments, the time required for full ecosystem recovery can be centuries.[8]

Despite this vulnerability, and even partly as a result of it, mountain life is highly diverse. In the Andes, for example, because of the intrinsic risk of agriculture in a complex environment, a farmer may plant up to 47 varieties of potatoes to exploit subtle differences in climates and soils. Indigenous Andean crops include amaranth, quinoa, cocoa, and more than 200 species of potatoes. And at the annual fair in Dali, in China's mountainous Yunnan province, thousands of Bai, Yi, Naxi, Tibetan, Lisu, and Lahu people trade as many as 550 species of native herbal medicinal plants in addition to hundreds of food plants.[9]

This natural diversity has also enabled mountain communities to become the custodians of vital crop species, helping to ensure genetic variety as well as disease and pest resistance. On the top of a 2,880-meter mountain in the Mexican Sierra de Manantlan, for example, can be found the only known stands of the most primitive wild relative of corn (or maize). Coffee, originally from the Ethiopian highlands, is now grown on tropical uplands around the world. And hundreds of years ago, the highland Quechua of Peru revealed to Europeans the anti-malarial medicine quinine. Mountain reserves of genetic diversity are critical insurance in a global economy that has converted many ecosystems to high-yielding monocultures vulnerable to pests, diseases, and changing soil and climate conditions.[10]

Mountain regions also often function as sanctuaries and refugia for plants and animals that have long since disappeared from lowlands transformed by glaciation or extensive human settlement. Freestanding mountains, such as Mount Kenya or the Tatra Mountains (shared by Poland and the Czech Republic), are islands of biodiversity in seas of humanly transformed landscapes. Andean countries are home to over half of all Neotropical bird species, 44 percent of the region's mammal species, and 38 percent of its amphibians. Similarly, some 3,000 plant species thrive in the region around the Chinese portion of Namche Barwa, in the eastern Himalaya. Within the numerous altitudinal belts of ecosystems in California's Sierra Nevada range are found an estimated 10,000 to 15,000 species of plants and animals.[11]

Mountains also rank high by another indicator of biodiversity: endemism, the occurrence of species only within narrow ranges, typically after centuries or more of isolation. Centers of endemism are areas where many such species occur together, and therefore are typically found on true islands or terrestrial island-like habitats such as isolated mountains. Seven of the world's 14 tropical "hotspots" of endemic plants threat-

Mountain regions often function as sanctuaries and refugia for plants and animals that have long since disappeared from lowlands.

ened by imminent destruction have at least half their area in mountains. In the United States, mountainous regions with high levels of endemism include the southern Appalachians, Ozarks, Klamath Mountains, and isolated ranges in the West and Southwest. Partly because of these high proportions of endemic species, mountain life is inordinately subject to extinction. For example, of all threatened mammal species in Australasia and the Americas, 19 percent are found in the montane tropical rain forests and an additional 12 percent in the alpine and other montane life zones.[12]

Birds are also excellent biological indicators because they occur in all regions, respond swiftly to environmental change, and are among the best monitored and understood creatures. Birds with restricted ranges tend to occur together in isolated patches of habitat, particularly montane forests. A study published in 1992 by BirdLife International revealed that 20 percent of the known 9,600 bird species were confined to just 2 percent of the earth's land surface, in pockets of endemism that also host 70 percent of the world's threatened birds and large numbers of other endemic animals and plants. Subsequent analysis revealed that 131 of the 247 endemic bird areas are in tropical mountains. The mountains of South America alone contain 40 of these—many of them confined to rare tropical montane cloud forests.[13]

Among conservationists from temperate countries, mountains have been somewhat misrepresented by the biological axiom that species diversity declines with increasing elevation. While this is generally true outside the tropics, numerous exceptions exist. For example, species of birds and butterflies tend to become more numerous above sea level. The diversity of lichens and bryophytes (mosses and liverworts) in temperate rain forests, found on western coastal mountains, appears to be comparable to the variety found in lowland tropical rain forests. And in the humid tropics, the number of epiphyte, shrub, herb, and fern species also increases with altitude. But what has largely escaped notice is that mountain environments are especially diverse at levels of biological organization higher than species—genera, families, phyla, habitats, and ecosystems. Perhaps no other life zone contains so much variation between habitats and ecosystems (known by biologists as "*beta* diversity") as mountains. The eastern slopes of the Andes are the paramount example of this variety. On these and other tropical mountains, vegetation belts may range from lowland tropical rain forest to moist submontane forest to dry or wet montane tropical forest, subalpine forest, alpine heaths, cloud forests, alpine grasslands, tundra, and permanent snow and ice fields. Each of these life zones—even the ice fields—has its own habitats and assemblages of plants and animals. In a recent study of global patterns of terrestrial plant diversity, Samuel Scheiner and José Rey-Benayas of Northern Illinois University found that the most diverse landscapes were areas with warm temperatures, high elevations, large areas, and large seasonal temperature fluctuations—all characteristics of tropical mountains.[14]

Because mountains tend to be challenging environments in which to produce a living, and often inaccessible, they have provided sanctuary to refugees, indigenous peoples, and ethnic minorities. Mountain people typically live on the economic margins as nomads, part-time hunters and foragers, traders, small farmers and herders, loggers, miners, wage workers, or in households headed by women while men pursue seasonal work elsewhere. Because men are largely involved in the cash economy, women in the mountains of developing countries often serve as the primary environmental managers and custodians of the local natural resources and biodiversity. Given the imperative to survive, mountain people have acquired unique knowledge and skills by adapting to the specific constraints and advantages of their fragile, inhospitable environments. They possess millennia of experience in shifting cultivation, terraced fields, medicinal use of native plants, migratory grazing, and sustainable harvesting of food, fodder, and fuel from forests. In India's Garhwal Himalaya, for example, local women recently were able to identify 145 species of plants that had been destroyed by commercial logging and limestone mining in the area, whereas national foresters could list only 25.[15]

Most of the 98 million people in China who are among the world's "absolute poor" live in mountains.

With human survival so closely dependent on knowledge of local ecology, the existence of sharp differences in the kinds of native foods, fuels, or medicines that can be found or grown in different parts of a single mountain range can result in similarly sharp differences in the knowledge—and hence the culture—of that range's human inhabitants. "Cultural diversity is not an historical accident. It is the direct outcome of the local people learning to live in harmony with the mountains' extraordinary biological diversity," says Anil Agarwal, founder and director of the Centre for Science and Environment in New Delhi. These two types of diversity—biological and cultural—are inextricably linked.[16]

Estimates of the number of the world's remaining indigenous peoples vary from 200 million to 600 million. Although no total figures exist on the numbers living in mountains, the proportion is clearly high, since mountains account for a substantial portion of the landscape that has not been transformed by modern economies. Indigenous peoples have learned to live on what nature offers. For instance, more than 10 million Quechua peasants, descendants of the Incas, now reside in the central Andes of Bolivia, Ecuador, and Peru. And more than 16 million indigenous farmers and pastoralists live throughout the 19 major mountain ranges of the former Soviet Union. Altogether, the world's mountains are home to several thousand different tribes or ethnic groups.[17]

For all their differences, however, one attribute most mountain peoples share in common—whether in the Ethiopian highlands, the Appalachians, the Andes, or even in the Swiss Alps—is material poverty. More than 60 percent of the rural Andean population lives in extreme poverty. Most of the 98 million people in China considered to be among the world's "absolute poor" are ethnic minorities who live in mountains. Similarly, residents of the Appalachians are among the poorest in the United States. (See Table 2.) Of course, while these data consistently indicate impoverishment, they should be interpreted with caution because such measurements fail to reflect the value of two common mountain village institutions—barter trade and common property resources such as forests and pastures. These data also fail to convey the cultural wealth and social integrity of many mountain communities, compared to their more transformed lowland counterparts. But notwithstanding these statistical shortcomings, the pressing need for more equitable development in mountains is clear and pervasive.[18]

TABLE 2

Human Development Indices, Selected Mountainous Countries[1] and Regional Averages

	Gross National Product per capita (1992 $)	Population Growth Rate, 1980-1992 (percent per year)	Life Expectancy, 1992 (years)	Infant Mortality, 1992 (per 1000 live births)	Adult Literacy, 1990 (percent)	Urban Population, 1992 (percent of total population)
Bolivia	680	2.5	60	82	77	52[a]
Ecuador	1,070	2.5	67	45	60	58
Guatemala	980	2.9	65	62	55	40
Haiti	370[a]	2.1[a]	55	94[a]	53	29[a]
Honduras	580	3.3	66	49	73	45
Peru	950	2.1	65	52	85	71
LATIN AMERICA AND THE CARIBBEAN	2,690	2.0	68	44	85	73
Burundi	210	2.8	48	106	50	6
Ethiopia	110	3.1	49	122	62[b]	13
Madagascar	230	2.9	51	93	80	25
Malawi	210	2.9	64	134	48[d]	12[a]
Rwanda	250	2.5	46	117	50	6
Tanzania	110	3.0	51	92	55[c]	22
SUB-SAHARAN AFRICA	530	3.0	52	99	50	29

[1]Includes high plateaus.

Source: World Bank, *World Development Report 1994: Infrastructure for Development* (New York: Oxford University Press, 1994), and on the following: [a] from World Bank, *World Development Report: Investing in Health* (New York: Oxford University Press, 1993); [b] (for 1980-85) from World Bank, *Social Indicators of Development* (Baltimore: Johns Hopkins University Press, 1994); [c] (for 1992) from United Nations Development Programme, *Human Development Report 1994* (New York: Oxford University Press, 1994); [d] (for 1987) from United Nations Educational, Scientific, and Cultural Organization, *Statistical Yearbook 1994* (Paris: 1994).

In view of the close link between mountain environments and cultures, it might seem puzzling that some of the world's most biologically diverse environments are homes to some of the planet's poorest people. The explanation lies partly in the geographical fragmentation of these regions; with their territories slivered by topographical and political boundaries, isolated mountain peoples have little or no voice in national affairs. The combination of political marginalization and poverty has subjected mountain peoples to widespread discrimination by dominant lowlanders, who refer to them with such pejoratives as "hillbillies" (United States), "oberwalder" (Austria), "kohestani" (Afghanistan), or "bhotias" (India). This discrimination can extend to violent human rights abuses—especially of ethnic minorities resisting political control by national governments. Prominent examples include Tibetans, Kashmiris, Kurds, Irian Jayans, and other minorities in Bhutan, Bosnia, Burundi, Colombia, Guatemala, Malawi, Peru, and Sri Lanka.[19]

These discriminatory attitudes reflect the persistence of a colonialist mentality towards mountain peoples, and towards the extraction of their natural resources for consumption primarily in the plains. Throughout history, mountain regions have been net exporters to the plains. Increasing movements of products, people, and information in both directions often facilitate political control and economic integration by linking once inaccessible, remote areas to market economies and central governments. Although this means better education and health services can reach the more accessible communities, it also invites cultural disruption, inequitable terms of trade, and sudden dependence on the lowland market economy. Relative to adjacent lowlands, per capita levels of investment tend to be far lower in mountainous areas. Mountain peoples usually have little voice in the nature of the transformation of their homelands, whether by a large hydropower dam, a ski resort, or a copper mine.[20]

People in these areas are further marginalized by their frequent position astride disputed political borders. Remote mountainous regions have often served as rugged redoubts from which ethnic groups resist or avoid control by their national governments. Not surprisingly, war and insurrection are common in the mountains. In 1993, of 34 major armed conflicts involving state forces taking place in 28 countries, 22 took place primarily in mountains, and another 8 included such areas. Since World War II, the mountains of Bosnia, Guatemala, Iraq, Iran, Laos, Kashmir, North and South Korea, Rwanda, Tibet, Vietnam, the Caucasus, the Ethiopian highlands, the Hindu Kush, and the Peruvian and Colombian Andes have been substantially damaged by military activities.[21]

Finally, beyond the collective assaults of environmental degradation, economic impoverishment, and social marginalization, mountain cultures are suffering numerous affronts to what more than 1 billion people consider to be sacred places—places that are the sites of innumerable temples, shrines, and other religious monuments. In mythology, mountains are often the "cosmic pillar" by which people or gods can ascend to or descend from the sky. Although increasing secularism, urbanization, and industrialization have distanced most industrialized societies from a sacred attachment to the land, mountains are still widely regarded as places of vital spiritual value. Examples of sacred mountains include Sinai and Zion in the Middle East, Olympus in Greece, Kailas in Tibet, T'ai Shan in China, Fuji in Japan, and the San Francisco Peaks in Arizona. Whether through religious, aesthetic, or recreational experiences, mountains help nourish the human spirit.[22]

In 1993, of 34 major armed conflicts in 28 countries, 22 took place primarily in mountains.

Through the movements of air, water, soil, animals, and people, mountains are integrally connected to the plains below them. These systemic connections comprise an interdependence that gives the high and steep places immense natural and cultural richness, but also renders them highly vulnerable to external forces.

Farming the Slopes

Time-tested mountain farming practices, many of which go back centuries, are being abruptly threatened by population growth, the fast-growing global economy, and overwhelming cultural influences from the plains. As roads and communications have improved accessibility to mountain communities, the traditional small-scale barter trade of modest farm surpluses has given way to trade in a few commercial crops demanded by mass markets. Often, the shift has been rapid, producing adverse economic, ecological, and cultural impacts. Where mountain ecosystems are fragile and cultures isolated, these impacts have been especially acute.

In many developing countries, development policies have actually undermined peasant agriculture rather than aiding it, and have left mountain farm communities enmeshed in intertwined webs of expanding population, declining resources, poverty, and environmental degradation. This degradation has become visible in several trends over the last half century: landslides have become larger and more frequent; water flows in traditional irrigation systems have fallen; and yields of major crops have not kept pace with the gains typically achieved in the plains. The genetic diversity of crops and livestock has been diminished, as has the diversity of flora in forests and pastures. The regenerative capability of the land, based on intricate linkages between various land uses, has been weakened. The periods of hunger between harvests have lengthened; more time is spent collecting fodder and fuel; and the rates of poverty, unemployment, and migration out of the hills have generally increased.[23]

Prior to colonial control, most mountain societies existed for generations as sophisticated agrarian societies. In delicately balanced relationships with their hazard-prone environments, they relied on complex strategies for deriving sustenance from their sloping patches of land. These now-threatened strategies are as diverse as the landscapes in which they occur; they range from long-fallow shifting agriculture (rotation cycles of several years) to intensive multi-cropping agriculture (two to three crops each year), and seasonal movements of livestock between lower and upper grasslands (known as "transhumance"). Key inputs to crops are typically self-generated: farmers store seeds from one harvest to the next, and fertilize their crops with animal dung. Detailed knowledge of how to manage and conserve natural resources is acquired only over several generations—comparable to the time mountain ecosystems require to respond and recover from major disturbances.[24]

Indigenous agriculture is generally aimed at sustaining the long-term productivity of the land for local consumption, not short-term maximization of yields for sale. Since mountains are often comprised of narrow altitudinal belts of complex ecosystems, successful mountain agriculture requires a much higher level of risk management than is required in less diverse lowland systems. For several centuries in the central Andes, farmers have employed the strategy of storing any one year's surplus as insurance against the possibility of the next year's shortfall. In another common strategy, farmers in Cuyo Cuyo, Peru plant up to 20 dispersed fields scattered through different microclimates. This region between La Paz and Cuzco, formerly the heart of the Tiahuanco and Incan empires, once supported more people in a state of relative sufficiency than occupy the area today in poverty. Some agronomists estimate that the abandoned Incan irrigated terraces and ridged fields could be restored to productive use at about one-tenth the cost of irrigating a comparable area in the arid coastal valleys, where substantial investments have been made to expand irrigation.[25]

Prior to colonial control, most mountain societies existed for generations as sophisticated agrarian societies.

As pressures mount to adopt commercial farming practices, however, such traditional methods of resource management are more likely to be lost than restored, and agriculture is usually intensified on marginally productive lands. Long fallow periods and rotations are shortened, and croplands are expanded. This forces farmers to pursue higher yields through increasing reliance on the high-productivity methods of commercial lowlands farming. These practices include the use of fewer seed strains with higher yields; heavier reliance on irrigation, credit, and other institutions of the market economy; and progressively higher doses of pesticides, herbicides, and fertilizers. Thus, crops like fruits and vegetables are grown to generate higher returns in the market economy. More traditional staple crops—such as barley, wheat, or rice—are abandoned. While some communities benefit financially by shifting from producing their own food to producing more profitable cash crops and livestock, they nonetheless become quickly vulnerable to the vicissitudes of the lowland market economy and its institutions—over which they exercise little, if any, control. Altogether, swift adoption of these farming practices developed in the plains—for plains conditions—can leave mountain farmers economically dependent and ecologically vulnerable.[26]

Another common trend in the shift to commercial agriculture is an increase in the size of livestock herds. Often this is accompanied by a shift from sheep and goats to cattle, which can generate higher profits. Without proper management, however, this more intensive use of grasslands often leads to depletion of highly nutritious and palatable grasses, invasions of noxious weeds and non-palatable grasses, and soil compaction. The land becomes less productive, with lower livestock carrying capacity. Ecological damages from intensive livestock operations include decline of fisheries as water is diverted for irrigation and stream habitats are degraded, diseases in native herbivores, and major changes in fire frequency, soils, and hydrology. Because livestock congregate in riparian ecosystems, which are among the biologically richest habitats in arid and semiarid lands, the ecological costs of grazing are magnified. Half of U.S. rangeland, much of it in the mountainous West, is now considered severely degraded, with its livestock carrying capacity reduced by at least 50 percent.[27]

A primary force driving these shifts in modern mountain agriculture is population growth. In the East African highlands of Burundi, Kenya, Rwanda, Tanzania, and Uganda, the population has mushroomed sevenfold in the last 80 years. Over the last few decades, growth rates in most Andean and Himalayan regions have averaged above 2.5 percent per year, equivalent to a doubling time of less than 28 years. Such demographic increases can place enormous pressures on marginal hillside lands. Tolerance for exceeding the human carrying capacity of mountain slopes is often narrow. In mountainous Rwanda, for example, although volatile ethnic and military rivalries certainly sparked the genocidal massacre of 1994, a powerful—and underreported—contributing cause was the country's acute scarcity of land: Rwandans had less than three one-hundredths of a hectare of arable land per person, equivalent to a square plot about 17 meters on each side, most of it on steep slopes. The use of lowland cultivation methods on some of those slopes has resulted in landslides that not only destroy property, infrastructure, and lives, but diminish for centuries to come what remains of the country's natural productive capacity. In Nepal, almost all the 55 hill districts are now unable to produce enough food to feed themselves.[28]

In the Andes, women produce, process, and sell up to 80 percent of their countries' food and run 70 percent of the small enterprises.

In the central Himalaya of India and Nepal, as well as in much of the Andes, another outcome of increased crowding on a declining resource base is outmigration—mostly of young men—to the plains for cash-producing jobs, especially in the off-season. In many mountainous regions, women now work significantly longer hours in the fields and forests than do women in the plains, and they do more of the work than men. In the Andes, women produce, process, and sell up to 80 percent of their countries' food and run 70 percent of the small enterprises; yet they have little to no access to land, credit, or technical support. A survey of several mountain communities in Nepal found that women of poor households performed 98 percent of the labor in the forests and 60 percent of the duties involving animal husbandry, agriculture, and water. The gender imbalance created by male departures causes a disproportionate impoverishment of females, along with increased drudgery for those who are left to tend the fields, care for livestock, and raise children. Often, increasing entanglement with the cash economy snatches away what little control women previously have exercised over such responsibilities as community forests (used for collecting fuel, fodder, and food) and grazing lands. As these communal lands are privatized, overused, or abandoned, women and the poor are affected most, since their access to the lost resources is usually not replaced with access to modern technology, credit, and other enabling links to commercial markets.[29]

A related force now pushing farmers off their ancestral slopes, especially in Latin America, is a pervasive pattern of inequitable distribution of land. Highly concentrated ownership of the more fertile and productive arable land in the plains creates pressure for growing populations to migrate and clear more marginal land on hillsides. In the mountainous countries of Guatemala, Ecuador, and Peru, nearly 90 percent of the farms are *minifundios,* often too small to provide an adequate income. At the other end of the spectrum, *latifundios* (large land holdings) control over 80 percent of the occupied land in Chile and 40 to 50 percent in Colombia, Ecuador, and Guatemala. While land reforms in the 1960s and 1970s provided some redress for millions of families, most of these reforms lacked coherent long-term strategies for sound resource management. Thus, lack of secure access to resources continues to drive peasants uphill onto less stable soils, or downhill to the cities or lowlands.[30]

Those who aren't driven from their land by population pressures or inequitable land distribution, or by the lure of higher incomes in the plains, can often find themselves economically enticed to engage in practices that are alien to their traditional farming culture—especially in remote mountain regions where law enforcement is scant or absent. Perhaps the most vivid illustration of the market economy's cultural invasion of traditional mountain economies is that of illicit drug production. By coincidence of the right soil and climate conditions and the poverty of peasant farmers, most of the world's cocaine, heroin, and opium production is concentrated in three compact mountain regions. Farmers in southeast Asia's "Golden Quadrangle" (northern Thailand, Myanmar (Burma), Laos, and southwestern China) and southwest Asia's Golden Crescent (northern Pakistan and Afghanistan) provide nearly all the poppies used in opium and heroin production. Similarly, most of the coca leaves used for cocaine and crack production are grown in the White Triangle of the Andean regions of Bolivia, Colombia, and Peru. Between 1979 and 1987, Peruvian cocaine producers cleared some 180,000 hectares, accounting for about one-tenth of the country's deforestation during that period. Although prices paid to farmers for these crops may prove irresistible, drug production and trade create environmental and cultural devastation—deforested hillsides, declining soil fertility, soil erosion, and water pollution, in addition to the ultimate harvest of drug addiction, AIDS, and violence among competing drug lords.[31]

As competition drives the production of ever more goods for the market, the erosion of age-old systems of managing common property—especially forests and pastures—worsens. The most advanced form of this social and economic deterioration can be seen throughout the Alps, where many mountain farms—on which fields have been cultivated for up to one thousand years— have been partially or completely abandoned as their owners become service workers at ski resorts or move downhill to the cities for work. Without regular cultivation and grazing— processes to which these modified hills have become ecologically adapted over centuries—their slopes become destabilized as they enter a transitional phase before returning to forest. During this transition, they are particularly prone to erosion; depending on the degree of slope, they can be swept by avalanches and mud slides. In 1992, the Swiss government passed a new law to subsidize the management of pastures and forests that are critical to protecting roads, railways, and settlements from uphill natural hazards, such as avalanches, floods, and rock slides.[32]

New croplands are often cleared at the ecological expense of forests vital to the stability of watersheds, or on steep, erodible lands better left fallow. Indicators of ecological imbalance include increased landscape instability and degradation, reduced biodiversity, and loss of crop cultivars and livestock breeds.

Throughout the Ethiopian Highlands, eroding hillsides show graphic evidence of the desperate poverty that drove recent migrants onto the most marginal, unstable slopes. During this century, intensive use by increasing numbers of people has gradually led the country's population to overwhelm its productive capacity—leading to degraded soils and vegetation over much of the country. With an annual population growth rate that has averaged 3 percent in recent decades, the pressures on agricultural lands have been devastating. On cultivated and fallow lands, annual soil losses now average 42 tons per hectare, more than twice the losses from nearby land in other uses. These losses have caused a reduction of 1 to 2 percent in annual crop production. Biological deterioration may add another 1 percent loss, since dung and crop residues are burned instead of being spread on fields as fertilizer.[33]

In the Himalaya, mountain farmers often have been blamed for causing massive downstream floods in the Ganges and Brahmaputra Rivers as a result of their upstream deforestation. Long-term scientific research, however, has revealed a much more complicated reality: in most traditional mountain farming systems, the contribution to erosion is negligible; but in newly settled mountain areas, migrants unfamiliar with local ecosystem processes sometimes apply agricultural practices better suited to more resilient lowland areas, and swiftly and irreversibly destroy the productive capacity of their hillsides. Thus, while cultivation of steep slopes can accelerate soil loss and thereby contribute to floods in small catchments, it is other factors that contribute most heavily to the massive sedimentation and flooding seen in large rivers. The main contributors are natural erosion following heavy rains in downstream catchments, erosion from commercial extractive industries (especially logging and mining), and poorly constructed road cuts. Nonetheless, much remains unknown about the precise dynamics of soil and water within most watersheds.[34]

Perhaps the most effective expression of the poor fit between modern development approaches and traditional mountain societies is the Chipko (literally meaning, "hug the tree") movement in the Garhwal Himalaya of northern India. In the early

1970s, villagers recognized that increasing local floods were due to widespread commercial clearing of hillside forests. Responding to an overall social and ecological disintegration of hill society over a few decades, village women launched a nonviolent peasant movement to preserve access to local forests. By all accounts, the movement was successful: in 1982, commercial logging of live trees was banned for 15 years in the eight hill districts of Uttar Pradesh in the Indian Himalaya.[35]

While a return to the virtual self-sufficiency of traditional subsistence agricultural systems is neither possible nor desired in most mountain communities, making mountain agriculture sustainable will also not be possible unless it includes a painstaking integration of indigenous knowledge with modern technologies and practices. Moreover, the modern inputs cannot be effective unless they are developed for—or adapted to—specific local ecological problems and human needs. And perhaps most important, new approaches to mountain agricultural development need to reinforce the versatile self-reliance of local communities by strengthening, not subverting, community traditions and institutions. With a long-term focus on meeting these needs, the productive capacity of mountain lands and the cultural integrity of their peoples can be restored.

Extracting Timber, Water, and Minerals

Farmers are not the only people having significant impacts on mountain environments. In most mountain countries, economic development policies have led to the expropriation of customary land rights and their redistribution to a range of vested commercial interests. Facilitated by the incursion of roads, bridges, and tunnels, these industries have transformed mountains into steep storehouses of timber, water, hydroelectricity, and minerals for export to the plains. Extractive industries, commercial operations, and large-scale interventions often cause exceptional ecological and cultural damage in mountains because they ignore the fragile ecologies and the rights of local communities. With half of humanity living downstream of mountain watersheds, the effects of this damage can be extensive.[36]

Forest destruction is moving up mountain slopes in most tropical countries. In a recent assessment of tropical forests, the U.N. Food and Agriculture Organization (FAO) found that during the last decade, tropical mountain forests have had the fastest rates of both annual population growth and deforestation. (See Table 3.) FAO concluded that hill and mountain forests were more subject to ecological damage from high population densities than all three types of lowland forest. The Guatemalan Highlands and the Bolivian Altiplano are two clear examples of the causal linkage between high population growth and severe deforestation. Deforestation is inherently a complex phenomenon, but in tropical mountains it is driven largely by population growth, uncertain land tenure, inequitable land distribution, and the absence of strong and stable institutions in remote regions. Although these pressures are common throughout tropical mountain forests, the resulting human dynamics differ. Throughout much of southeast Asia and China, colonists from crowded lowlands have pushed upland farmers, whose land tenure is often insecure, ever higher into the mountain forests. The uplanders, in turn, have often encroached on the homelands of indigenous peoples. Conversely, centuries of population pressure and intensive land use in the Andes and East African Highlands have diminished those forests to small remnants of what they once were, and have pushed mountain people downhill to the less productive lowlands.[37]

Tropical montane cloud forests provide good illustrations both of the unique assemblages of vegetation found in tropical mountains and of the impacts people can have on them. These unique forests occur where there are persistent, seasonal, or frequent wind-driven clouds from which the forest can strip or "harvest" atmospheric moisture above and beyond normal rainfall. Found most commonly on tropical and sub-tropical mountains subject to oceanic climates, this horizontal (or "occult") precipitation is intercepted on the ample surface areas of trees, shrubs, epiphytes, mosses, and lichens. Typically, it amounts to 5 to 20 percent of ordinary rainfall, or hundreds of millime-

TABLE 3

Tropical Population, Forest Cover, and Forest Loss, 1981-1990

Primary Forest Ecosystem Type	Population Density, 1990	Forest Cover, 1990	Population Growth, 1981-90	Forest Loss, 1981-90
	(people per square kilometer)	(million hectares)	(percent annual change)	(percent annual change)
Hill and Mountain	56	204.3	2.6	1.1
Lowland Rainforest	41	718.3	2.2	0.6
Lowland Moist Deciduous	55	587.3	2.4	1.0
Lowland Dry and Very Dry	70	238.3	2.3	0.9

Source: Adapted from Food and Agriculture Organization (FAO) of the United Nations, Forest Resources Assessment 1990: Tropical Countries, FAO Forestry Paper 112 (Rome: 1993).

ters. When these forests are cleared, however, this extra harvest of water is lost or greatly reduced—along with the critical roles all forested headwaters play in maintaining water quality and stabilizing water flows and hillsides. Cloud forest ecosystems also host a high proportion of endemic species. For instance, more than 21 percent of the world's restricted-range bird species have cloud forest habitat.[38]

As recently as the early 1970s, cloud forests covered up to 50 million hectares in narrow mountainside belts. But these forests have been under severe pressure, disappearing faster than the better-publicized lowland tropical rain forests. Neotropical botanists estimate that almost 90 percent of these forests have been lost in the northern Andes—largely to the expansion of grazing from both above and below. Worldwide, many cloud forests are ecologically degraded by unsustainable fuelwood and charcoal wood cutting, as well as commercial extraction and trade in their plant and animal life, including orchids, bromeliads, birds, amphibians, and medicinal plants. Clinging to remnant patches of these forests are mountain gorillas in central East Africa, spectacled bears in the Andes, and resplendent quetzals in Central America.[39]

In the temperate zone, a similarly rare and rich type of forest has met a comparable fate. Temperate rain forests occur on the slopes of western coastal mountains that receive abundant rainfall year-round. No other kind of terrestrial ecosystem produces as much living matter (biomass) per unit of area—as much as 500 to 2,000 tons per hectare. Some of the world's oldest and largest trees grow in these forests. Once covering 30 to 40 million hectares, they now span about 14 million hectares in the North American Pacific Northwest, Chile, Tasmania, and southern New Zealand. The global demand for the wood products of these forests has been the primary force driving their destruction. Among the most valuable timber species are South America's alerce and monkey puzzle and North America's Sitka spruce, yellow cypress (cedar), and Douglas fir. After more than a century of logging in Washington and Oregon, only one large watershed remains entirely unlogged. Since 1950, more than one-half of the most productive rain forest in Alaska's Tongass National Forest has been clear-cut. In Chile and New Zealand, much of the drier hardwood forest has been cleared for exotic pine plantations.[40]

Several practices of commercial forestry—burning slash, skidding, the introduction of exotic plants, and the spraying of herbicides and pesticides—are likely to be more destabilizing to mountain slopes than to flatter lands. Altogether, these industrial forestry practices tend to erode hillsides, reduce the structural and species diversity of the forest, and impair the natural processes of forest ecosystems. The western side of California's Sierra Nevada range, for example, illustrates the legacy of industrial forestry in mountains. In the last quarter century, more than 100,000 hectares of low- to mid-elevation forest have been clearcut. On the steep sections where soils are usually unstable, soil erosion rates exceed those of soil formation by 20 to 40 times. This erosion often proves harmful to fish and other aquatic species. Adding to these damages are 30,000 kilometers of roads that have allowed farming, ranching, mining, and

human settlement—in addition to logging—to spread throughout most of the region. The remaining unprotected wilderness
(i.e., roadless, intact ecosystems), covering about 500,000
hectares, is located mainly on the steepest and least accessible terrain with the youngest soils and least productive vegetation.
Nor is this situation unique to the
Sierra Nevada. On Forest Service
lands throughout the United States,
a substantial portion of the remaining mature and old-growth forests is
found on mountain slopes.[41]

> **Cloud forests have been disappearing faster than the better-publicized lowland tropical rain forests.**

Because their elevation and slope
permit gravity to increase the force of
flowing waters, mountains attract
most of the world's hydroelectric power projects, and many irrigation reservoirs as well. In the Swiss Alps, almost every possible site for a hydroelectric power facility has been developed.
Many mountain river systems in the United States—including
those of the Tennessee, Colorado, and Columbia Rivers—have
been converted into stepped sequences of linked reservoirs. In
the continental United States, Norway, New Zealand, and
Mexico, most of the commercial hydropower potential already
has been developed.[42]

Most developing countries with major hydropower potential,
on the other hand, are actively pursuing big projects to meet substantial shortfalls in electricity supplies, despite frequent local or
international opposition to the high economic, social, and environmental costs that many of these megaprojects incur. As "a
showpiece of industrial progress," Malaysia is spending $5.8 bil-
lion to build the 2,400-megawatt Bakun dam in Sarawak state.
This project will require construction of a 210-meter-high dam,
the relocation of more than 8,000 tribal people, inundation of
69,500 hectares (an area larger than Singapore), and the clearing
of 80,000 hectares of mostly intact moist tropical forest for 670
kilometers of power lines.[43]

On the mountainous upper reaches of the Yangtze River,
China's Three Gorges dam will be the world's largest hydroelectric project, at 18,000 megawatts. If ever completed, it will

also uproot 1.25 million people, flood 110,000 hectares, and cost at least $34 billion. Near Tehri in the Garhwal Himalaya, the Indian government is building a 260-meter-high hydropower dam, just a few kilometers away from where an earthquake killed 2,200 people in 1991. This project is being undertaken in the face of repeated protests by 105,000 mountain villagers who face forced relocation and by numerous seismic experts.[44]

The rocky Kingdom of Lesotho is engaged in a 30-year project involving five dams, two hydropower stations, and nearly 200 kilometers of underground tunnels to redirect the Senqu River and its tributaries under the Maluti mountains and into South Africa's industrial, mining, and metropolitan heartland, known as the Vaal Triangle. The 172-meter-high Katse dam will submerge 35 square kilometers of a steep valley, including 600 hectares of cropland and 2,600 hectares of grazing slopes. The already highly eroded project area suggests that sedimentation rates into the reservoir could be high. The project's $2 billion price tag is nearly four times Lesotho's gross national product.[45]

At the World Bank's celebration of its fiftieth anniversary in Madrid in October, 1994, more than 2,000 nongovernmental organizations from 44 countries called on the Bank to stop funding large dams. The NGOs charged that the $54 billion the Bank has invested in building large dams has harmed local peoples and the environment, while saddling recipient countries with excessive debt. The World Bank's International Finance Corporation is providing the lead financing for the $125 million Pangue dam on Chile's Bio Bio River, home to the Pehuenche people—one of Chile's five major indigenous groups. Together with two other planned dams, the project would flood or deforest more than half of Pehuenche territory. Litigation by Chilean NGOs over the project's downstream impact has failed to halt the project.[46]

Of all the economic activities in the world's mountains, nothing rivals the destructive power of mining. Environmental impacts include habitat destruction, increased erosion, air pollution, acid drainage, and metal contamination of water bodies. In Colorado, for instance, when Galactic Resources of Canada declared bankruptcy in 1992 and abandoned its gold mine in

Summitville, the mine continued to leach cyanide, sulfuric acid, and toxic heavy metals into 28 kilometers of the Alamosa River. Summitville is now a Superfund hazardous waste site that will cost $100 million to clean up. The largest Superfund toxic waste site stretches 220 kilometers across Montana's Silver Bow Creek and Clark Fork River. More than a century of mining in the Clark Fork Basin has created 40 billion liters of acid mine runoff that spreads outward each year, threatening aquifers that flow into several rivers on which millions of Americans depend.[47]

Of all the economic activities in the world's mountains, nothing rivals the destructive power of mining.

Now that the Andean countries have rewritten their mining codes to encourage foreign investment, dozens of multinational companies from Australia, Canada, South Africa, and the United States are rushing to extract the Andes' rich deposits of copper, silver, and gold. In the Khaniara area of India's Himachal Pradesh, nearly 1,000 small- to medium-sized slate mines have stripped up to 60 percent of the forest cover and triggered countless landslides.[48]

The Ok Tedi open-pit copper and gold mine in the Star Mountains of western Papua New Guinea is the country's second largest mine, and a vital source of income. In 1991, the mine shipped 600,000 tons of copper concentrates to Japanese smelters. This remote area is also home to the Wopkaimin people, for whom the mountain is sacred. By the time the mine closes, the 2,330-meter mountain will have been virtually leveled. After the failure in 1984 of a partially-built tailings dam, the mining company received government permission to dump 80,000 tons of tailings each day directly into the Ok Tedi River, which flows directly into the 1,100 kilometer-long Fly River. These tailings are laced with copper, cyanide, and heavy metals. The mine's toxic tailings have destroyed all the river's species of crocodiles, turtles, and fish—as well as all the riverside gardens that its rich sediments supported. The river water is unfit for drinking or even for washing clothes. As the river bed has risen five meters in ten years, large areas of floodplain forest have been destroyed, along with the birds and animals that once lived there. As the dumping has continued, 30,000 villagers recently filed a $3-billion class action suit against the company for pollution of the river and damages to its fish stocks and the adjacent fields.[49]

The impacts of extractive industries on the world's mountains are inexorably linked to the growth of the world economy, which has expanded fivefold since 1950. In that period, global per capita consumption of copper, energy, meat, steel, and timber has approximately doubled. To reduce the impacts of natural resource extraction in mountains will require systemic reforms at all economic levels, from local communities to the world itself.[50]

Playgrounds on High

In 1898, John Muir, a founder of the U.S. conservation movement, wrote, "Thousands of tired, nerve-shaken, overcivilized people are beginning to find that going to the mountains is going home; that wilderness is a necessity; and that mountain parks and preservations are useful not only as fountains of timber and irrigating rivers, but as fountains of life." Today, so many lowlanders are confirming Muir's observation that the problems they had hoped to leave behind are following them up the slopes.[51]

Explorers and climbers were the forerunners of today's recreational pilgrimage to the world's mountains. While they usually posed little threat to mountain ecosystems and cultures, their travel accounts have encouraged millions of tourists and trekkers to follow in their footsteps. In slick magazine advertisements and television commercials, images of pristine wilderness and primeval solitude lure people from the plains to the mountains for respite and sport. In industrial countries, mass tourism and recreation are now fast overtaking the extractive economy as the largest threat to mountain communities and environments. Since 1945, visits to the 10 most popular mountainous national parks in the United States have increased twelvefold. (See Figure 1.) The 9 million people who visit the Great Smoky

FIGURE 1

Recreation Visits to Ten Mountainous U.S. National Parks, 1915–1993

Million Visits

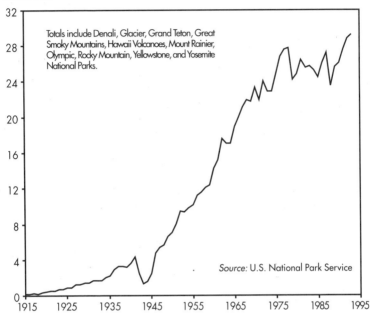

Totals include Denali, Glacier, Grand Teton, Great Smoky Mountains, Hawaii Volcanoes, Mount Rainier, Olympic, Rocky Mountain, Yellowstone, and Yosemite National Parks.

Source: U.S. National Park Service

Mountains National Park each year place immense pressure on its wealth of species, which include 1,400 flowering plants, 400 mosses, 156 birds, and over 2,000 fruiting fungi. In all its forms, tourism is now the fastest-growing industry in the western United States, and is the largest private employer in 7 of the 11 western states. The hundreds of millions of people who visit mountain ecosystems each year create an enormous collective impact, including non-biodegradable litter, trampled vegetation, and erosion from impromptu trails. Even worse, this tourism frequently induces poorly planned, land-intensive development.[52]

Tourists overwhelm mountain areas in the tropics as well. Every year in the Indian Himalaya, for example, more than 250,000 Hindu pilgrims, 25,000 trekkers, and 75 mountaineering expeditions climb to the sacred source of the Ganges River, the Gangotri Glacier. They deplete local forests for firewood, trample riparian habitat, and strew litter. While tourism can bring significant revenues to the local and national economy, it often places immense pressures on fragile ecosystems and isolated cultures. By creating new demands for local commodities, tourism can bring swift changes to traditional communities, including intensive cash-cropping, excessive fuelwood cutting, and a dramatic increase in livestock numbers. For instance, before the Barun Valley in eastern Nepal was set aside as a protected area in 1992, yak herders had cleared some alpine pastures to plant vegetables for sale to mountaineering expeditions, and had begun widespread burning of ancient juniper woodlands. The dead juniper scrub was then light enough to carry as firewood to the Makalu base camp, where it was sold for exorbitant prices. Nepal now has the world's fastest growth rate in tourism, with trekking permits alone increasing by 17 percent in 1993.[53]

In the European Alps, tourism now exceeds 100 million visitor-days, bringing $52 billion to the region each year and supporting 250,000 jobs. Much of this economic and demographic growth is concentrated in the western part of the eastern Alps. While industrialized zones and centers of tourism have boomed in that area, most of those adjacent valleys that are still primarily agricultural and inaccessible to tourism are losing residents—sometimes leading to the destabilization of their abandoned sloping fields and pastures. A recent demographic analysis of the European Alps found that over the last century most of the region's population growth occurred below the altitude of 500 meters, along the borders of the range and in the broad inner-Alpine valleys. Most communities above 1,500 meters had lost population.[54]

Leisure and recreation in the mountains can approach the absurd. In 1990, there were 100 golf courses in the European Alps; by 1992 there were 250; and by the end of 1996, 500 are expected. And the lure of golf in the mountains is not restricted to rich countries; tropical montane cloud forests in Malaysia's Mount Kinabalu and Cameron and Gentina Highlands are now being cleared for golf courses, and China is building a golf course and luxury hotel near the remote Yulongxue Shan in northwestern Yunnan. Usually built upon cleared forests or filled-in wetlands, golf courses require exceptional amounts not only of water, but also of insecticides, herbicides, and fertilizers that can harm birds, downstream vegetation, and fish. In Malaysia, one medium-sized course requires enough water each year to supply the needs of 20,000 people. Golf course construction is now the fastest growing type of land development in the world.[55] The impacts of tourists are not limited to places where bulldozers and builders can precede them. Helicopters transport well-heeled skiers to untracked slopes not only in the North American Rockies and European Alps, but also in the Indian Himalaya, Chilean Andes, Japanese Alps, Polish Tatra, New Zealand's Southern Alps, and Alaska's Pacific Coastal Range. In 1993, U.S. ski resorts alone hosted more than 55 million skier visits (a skier visit being defined as one skier for any part of one day). To expand or just maintain their business, many ski resorts now consume substantial amounts of water to make snow at the very time of the year when mountain water is least available. Colorado ski resorts now divert about 2,000 acre feet (one acre foot is 1.23 million liters) of water for making snow—three to four times the volume of water used a decade ago. While that amount pales in comparison to the volume used for irrigation, the snowmakers' demands often concentrate on small streams when flows are lowest in the late fall, increasing the risk that the streams will freeze and rip fish eggs from the gravel of stream beds before they can hatch in the spring. Since ski areas are usually cleared from mature forest and reseeded with exotic grasses, they permanently fragment forest habitat and often lead to soil erosion from cleared areas.[56]

In 1990, there were 100 golf courses in the European Alps; by the end of 1996 there are expected to be 500.

The populations of many small ski towns—like Vail, Colorado, and Park City, Utah—have more than doubled since 1980, causing the rates of new home construction and retail sales to increase at double or even triple the national average. As an extreme example of the results of such growth in a compact mountain valley, the median price for a single family home in the cosmopolitan ski town of Aspen, Colorado (population 5,046) now exceeds $1.5 million. These lofty enclaves of the recreational elite often create economic gradients that force wage workers in the service economy to migrate downhill from the resorts, leading to sprawling, hasty development in the surrounding region. Service workers now commute to Aspen from as far away as 140 kilometers (87 miles).[57]

The Rockies also have attracted large numbers of information-based businesses, as the proliferation of fax machines, modems, and other telecommunications technology has allowed office location to become more a matter of choice than of necessity. As more people have sought to escape crime-ridden, congested, and expensive coastal cities such as Los Angeles, places like Colorado and Utah—with their lower business costs, able work forces, pro-business culture, and outdoor recreation opportunities associated with the new mountain life-style—have attracted a growing stream of newcomers. The Rockies accounted for seven of the 10 leading U.S. states in new business incorporation in 1993. In the wake of this migration have come the latest versions of suburban sprawl: 16 hectares (40 acres) per residential lot in many neighborhoods, featuring "trophy" homes with more than 280 square meters (about 3,000 square feet) of floor space.[58]

The relocation of service industries is reshaping the economic landscape in many mountainous regions of industrialized countries. For example, a recent Wilderness Society study of the Yellowstone region in the United States found that during the last 20 years, 96 percent of the new jobs and 89 percent of the growth in labor income occurred in sectors outside agriculture, logging, and mining—the three legs of the traditional rural economy in the western United States. The study also found that oil and gas, mining, timber, and grazing in the ecosystem's seven national forests accounted for just 5 percent of the area's total employment, while recreation generated the majority of direct jobs in all but one of these forests.[59]

As retirees, entrepreneurs, and outdoor adventurers scramble to claim the last available parcels of mountain wilderness for their own, temperate mountain ranges are becoming new symbols of the "good life" for the affluent in the information-age and consumer society. Unfortunately, the good life of the high enclaves is demonstrably undermining the sustainability of all mountain life.

Trouble in the Air

Another dimension of the vulnerability of mountain environments, stemming directly from their distinctive vertical dimension, is the air pollution that these high landmasses intercept from industrial centers. In both the Appalachians and the Alps, proximity to industrial-urban corridors has proved damaging, or even deadly, to mountain forests. Since 1950, automotive use in Europe has increased fifteenfold in terms of vehicle-miles—driving up carbon monoxide emissions fivefold, hydrocarbon emissions sevenfold, and nitrous oxide emissions nineteenfold. Each year, at least 150 million people drive across the Alps, and this traffic is expected to increase by 50 percent or more by 2010. At the urging of those living near Switzerland's St. Gotthard pass, voters decided in 1994 that all heavy trucks passing through the country must move on railroad flatbeds by 2004. One of the most extreme results of increased pollution can be seen along the Czech-Polish border, where most of the forests of the Jizera and Giant Mountains have been damaged or destroyed by acid rain produced by emissions from automotive tailpipes and industrial smokestacks.

Each year, at least 150 million people drive across the Alps.

Damage is greatest in the highest spruce forests, where airborne acids concentrate in the clouds and on the needles of trees. Those at upper tree line are particularly affected because the airborne toxins concentrate just below the inversions that often form at this altitude. Similarly, in the industrial heartland of the United States, both the Great Smoky Mountains and the Appalachians experience harmfully high levels of acid precipitation and air pollution.[60]

Over the longer term, an even larger threat to the world's mountain environments may come from airborne emissions via their effect on the global climate. As greenhouse gas emissions have climbed, the melting of small glaciers throughout most of the world's mountain ranges in this century has provided tangible evidence of incipient climate change. Since they are partially composed of melting water, mountain glaciers are exceptionally sensitive to slight changes in temperature. A recent analysis of the historical records of small mountain glaciers around the world revealed an average warming of about 0.66 degrees Celsius over the last century. (See Table 4.)[61]

Although some climatologists question whether this observed warming is indeed the initial stage of a warming induced by human-caused emissions of greenhouse gases, or simply a phase of the planet's natural temperature fluctuation, climate change particularly threatens mountain ecosystems. Because mountains usually have steep temperature gradients over short distances, many of their plants and animals are particularly sensitive to such change. Climate scientists currently project an average warming of 1.5 to 4.5 degrees Celsius by 2050. A warming of 3 degrees Celsius would be roughly equivalent to an ecological shift upwards of about 500 meters altitude. Species already confined to the tops of mountains or below impassable barriers like rock outcroppings or highways could be exterminated as they are ecologically squeezed out of their habitat.[62]

In a survey of 26 summits, Austrian researchers recently found that as mean annual temperatures had increased by 0.7 degrees Celsius in this century, nine plant species typical to the nival life zone (above closed alpine grasslands) had migrated upward at an average rate of 1 meter per decade. Given an average drop in temperature of about 0.5 degrees Celsius for every 100 meters increase in altitude, the observed warming theoretically should have led to a shift in altitudinal vegetation of 8 to 10 meters per decade. The Austrian study, for the first time, confirms that the actual rate at which plants can migrate is slower than would be required to adapt to recent and anticipated warming.[63]

TABLE 4
Mountain Glacier Retreat, Selected Regions

Region	Number of Glaciers	Period of Observation	Scaled Mean Trend[1] (meters per year)
Alps	4	1850-1988	-9.3
Central Asia	9	1874-1980	-13.3
Iceland	1	1850-1965	-6.3
Irian Jaya	2	1936-1990	-7.1
Kenya	2	1893-1987	-6.7
New Zealand	1	1894-1990	-13.9
Norway	2	1850-1990	-12.1
Rocky Mountains	24	1890-1974	-13.7
Spitsbergen	3	1906-1990	-14.9

[1]The records for different glaciers were made comparable by a two-step scaling procedure that compensated for differences in glacier geometry and in climate sensitivity.

Source: Adapted from Johannes Oerlemans, "Quantifying Global Warming from the Retreat of Glaciers," Science, April 8, 1994.

The prospects for climate change causing rapid ecological changes in mountains extend far beyond rare species clinging to mountain tops. Mountain cloud forests are particularly vulnerable to climate change, since these relatively narrow altitudinal belts of vegetation are so specifically determined by climate. And in the vast Tibetan Plateau, according to one recent computer simulation, an increase of even 2 degrees Celsius in annual mean temperature would cause most of the current ecosystems to disappear and, in the central and northern sections, to be replaced with desert. Since increases in air pollution levels and climatic temperature would coincide with habitat degradation and other ongoing anthropogenic stresses on biodiversity, the cumulative impacts on mountain biodiversity could become devastating.[64]

Finally, most computer models of future climate change predict significant regional shifts in the timing and amount of precipitation. Beyond such local effects as some ski resorts facing the loss of sufficient snowfall for the financially critical holiday season, economic impacts could extend far downstream to a disruption of water supplies for irrigation, for downstream subsistence societies in the developing world, and even for entire cities—such as Los Angeles, Lima, and Cairo—that are dependent on mountain water for their existence. These possible shifts in the seasonality and quantity of fresh water flowing from mountain watersheds make the conservation of their ecological integrity all the more vital.[65]

Beyond Parks: Integrating Conservation and Development

More than a century ago in New Zealand, Maori people feared exploitation of sacred peaks by European sheep farmers and other colonists. They devised an ingenious solution: the Tongariro Mountains were given to Queen Victoria, and New Zealand's first national park was created. Globally, approximately 8 percent of all mountains are in some form of protected area, equivalent to the size of the U.S. state of Alaska and the Canadian province of British Columbia. But this total figure distorts the actual level of protection because it includes the 97 million-hectare Greenland National Park. On average, mountain protected areas are larger than those found in the lowlands, but there are wide variations in the degree of protection. For example, only 2.6 percent of the Alps is protected. By contrast, 70 percent of New Zealand's Southern Alps is protected, making them the most fully protected range in the world. Although mountain reserves occur on every continent, at every altitude, and in every biogeographic realm, they are far from evenly distributed. (See Table 5.) Underrepresented areas include the Alps, the Atlas range, the Hindu Kush, Papua New Guinea, the Sredinnyi Ridge in Kamchatka, and the mountains of Laos, Myanmar

TABLE 5

Inventory of the World's Mountain Protected Areas, 1992

Region	Biogeographic Realm[1]	Number	Total Area (hectares)
Antarctica and New Zealand	Antarctic	11	2,086,861
Australia	Australia	3	2,649,148
Europe, North Asia, North Africa, and Middle East	Palearctic	147	30,270,611
Latin America and the Caribbean	Neotropical	90	30,969,739
North America[2]	Neoarctic	96	153,804,175
Pacific	Oceania	8	3,598,032
South and Southeast Asia	Indomalayan	52	8,794,398
Sub-Saharan Africa[3]	Afrotropical	35	10,986,512
World Total		442	243,159,476

[1]Classification of world's biomes from M.D.F. Udvardy, *A Classification of the Biogeographical Provinces of the World,* IUCN Occasional Paper 18 (Gland, Switzerland: 1975). [2]Includes Greenland National Park, which has an area of 97 million hectares. [3]Includes Atlas mountains.

Source: Adapted from James W. Thorsell and Jeremy Harrison, "National Parks and Nature Reserves in Mountain Environments of the World," *GeoJournal*, May 1992.

(Burma), and Vietnam.[66]

With years of slashed budgets, national parks are straining just to provide the services and recreation demanded by visitors. Preservation of untrammeled wilderness has become increasingly secondary. Especially in developing countries, mountain protected areas often lack the necessary administrative and legal resources to provide effective protection to the biodiversity—genes, species, ecosystems, and landscapes—under their stewardship. In addition, many park designs are "doughnut holes," including only the rocky promontories above treeline, with their dramatic vistas—while much more biologically rich habitats in the slopes or valleys below them are leased for natural resource extraction or sold to private landowners. Others fail to protect the habitats they do contain. In the mountainous wilderness of the southern Canadian Rockies, for example, grizzly bear populations are declining because their habitat is being squeezed from within by millions of visitors and from outside park boundaries by expanding oil and gas production.[67]

Where endemic species are at risk, failure to protect the single site or the small number of localities where a species occurs can lead to rapid extinction. Some years ago, an isolated ridge in Ecuador called Centinela was found to contain at least 38, and possibly as many as 90, new species of plants—all but a few unique to the area. The forest on the ridge was destroyed by loggers, and most of the species are now extinct. Such losses can have cascading impacts. A recent study by the Centre for Population Biology at Imperial College in the United Kingdom found that as biodiversity declines, the performance of ecosystems—including the productivity of forests or grasslands, buffering against ecological disturbances, and the ability to fix carbon dioxide—also declines. Similarly, a long-term study of grasslands by ecologists David Tilman and John Downing found that prairie communities high in species diversity resisted drought and other stresses better and recovered faster than comparable communities with low species diversity.[68]

Many mountain species are especially vulnerable to harm from alien (non-native) organisms, since many mountains are essentially island habitats with no evolutionary defenses against invading species. The invaders can be brought in by human visitors or by the introduction of crops or ornamental plants. Usually imported without the predators or pests with which they evolved, these invasive species harm native species by outcompeting native flora and fauna. Examples of especially damaging alien species include feral pigs in Hawaii and Costa Rica, goats in Venezuela, and foreign grasses in Puerto Rico. Even Yellowstone Lake, in the heart of the oldest national park in the United States, is grappling with an alien trout species that threatens to wipe out the native cutthroat trout population. Eradicating alien species is time-consuming, experimental, and expensive.[69]

Another model for protecting mountain lands is the Man and the Biosphere (MAB) program launched by the United Nations Educational, Scientific, and Cultural Organization in the early 1970s. This model is more compatible than national parks (which typically forbid human habitation) with the populous, subsistence settlements often found in tropical mountains. Worldwide, more than 40 percent of the current roster of 324 biosphere reserves are located in mountains. Biosphere reserves are comprised of three geographic zones. A core zone serves as the strictly protected area, and is often a pre-existing park or preserve; surrounding the core zone is a buffer zone which allows such low-impact activities as education and training, non-intrusive research, and traditional subsistence agriculture and renewable resource extraction. Beyond the buffer zone is a transition zone for human settlements and a wide range of economic activities.

Many park designs are "doughnut holes," including only the higher zones above treeline and excluding the biologically richer lower flanks.

Unfortunately, this kind of zoning has seldom been implemented: most biosphere reserves have been superimposed on existing national parks and other protected areas without creating buffer zones that are critical for perpetuating wide-ranging species, disturbance regimes, and evolutionary processes. Thus, the ideal of the biosphere reserve remains largely unfulfilled.[70]

With so many pressures on mountain protected areas, conserving biodiversity will require much more than parks and biosphere reserves. When Aldo Leopold wrote an essay on "Thinking Like a Mountain" in his classic *A Sand County Almanac*, he was commenting on society's failure to understand the ecological balance maintained by predator-prey relationships in natural ecosystems. Leopold made a plea for finding "peace in our time" through cooperative management of protected areas and other natural resources. As competition and conflicts increase over ever scarcer mountain resources, conserving the full spectrum of biological variety will fail if patches of protected and unprotected land are managed in isolation from each other, without a comprehensive strategy. Leopold's injunction to "think like a mountain" points directly to cooperative projects with local mountain peoples as the institutional basis for integrating economic vitality and ecological integrity.[71]

Several nongovernmental organizations are pioneering approaches that combine biological conservation with equity, empowerment, and indigenous rights. These projects often return control of local mountain forests, fields, and ranges to local communities, while giving them the incentives and means to achieve resource practices that are sustainable. By valuing the tremendous diversity, limited production scale, and vulnerability of mountain environments, these community-based initiatives have established viable models for mountain peoples and organizations everywhere. (See Table 6.)[72]

One of these projects was set up in the Annapurna region of central Nepal—a region that stretches from subtropical lowlands and lush temperate rhododendron forests in the south to dry alpine steppe in the north. Its variegated environments include 1,422 plant species, 474 species of birds, 101 mammal species, and 100 species of orchids. Its natural diversity is rivaled by its cultural diversity: 116,000 people belonging to 14 distinct ethnic groups inhabit the 762,900-hectare area. The scenic grandeur belies an inherent vulnerability, however. More than 45,000 visiting trekkers ascend the trails every year, accompanied by a similar number of porters. The local population is growing by 2.8 percent a year, meaning it would double in just 25 years at current growth rates. More than 90 percent of the residents are subsistence farmers who depend on the depleted forests for food, fuel, fodder, and timber. Overgrazing by livestock and cultivation of crops on marginal lands add to hillside instabilities, allowing soil to wash away at an annual rate of 20 to 50 tons per hectare.[73]

In response to these ecological and economic threats, the King Mahendra Trust for Nature Conservation (KMTNC) founded the Annapurna Conservation Area Project in 1986. The project began with the villagers' priorities: a clean water supply

TABLE 6

Conservation and Development Projects in Mountain Communities, Selected Examples

Project/Program, and Organization	Location	Activities and Accomplishments
Makalu-Barun Conservation Project, *The Mountain Institute and Nepal's National Parks and Wildlife Conservation Department*	Makalu-Barun region, eastern Nepal	In a region of intact forests and high biodiversity, created 83,000 hectare conservation area around Makalu-Barun National Park for the 32,000 residents of seven distinct hill tribes; created 13 skills training programs and 10 cultural conservation projects; preserved some of Nepal's last riverine tropical forest; established 33 community forest user groups that manage 2,000 hectares of forests, two nurseries that can produce 60,000 seedlings each year, and kerosene depots at the trailhead to Makalu base-camp; project model now being replicated with technical and cultural modifications in Bolivian and Peruvian parks.
Hill Area Development Foundation	Chiang Rai province, northern Thailand	In heavily deforested watersheds, works with 28 villages of four tribal groups to build terraces, plant and rotate indigenous crop species along contours, form community forests, teach literacy, and help secure land tenure.
Mattole Restoration Council	Mattole River Valley, northern California	To reverse effects of soil erosion (produced by logging and overgrazing) on salmon and trout spawning, coalition of 100 community groups has planted thousands of native trees to control erosion; raised and released 250,000 native salmon since 1980 to help restore fisheries.
Bauda-Bahunipati Family Welfare Project, *World Neighbors*	Sindhupal-chowk District, eastern Nepal	Project nursery producing 15,000 fodder, fuel, and timber seedlings a year; family planning adopted by 22 percent of fertile couples, and fertility rate reduced from 5.8 to 3.2 children per couple; built 55 new drinking water systems and 525 pit latrines; project now replicated in 38 villages of 153,000 people and run by local NGOs.
Integrated Family and Communal Gardening Project, *AIDESEP*[1]	Peruvian Amazon	In response to abandoned farms, low productivity cattle pastures, dwindling territories, and assailed cultures, project provides training in organic crop production to 120 communities of 36 indigenous organizations; soil restoration has had 90 percent success rate, reducing toxicity from pesticides by 70 percent; supported 39 model gardens; now studying system for alternative land use model for granting communal land titles.

[1]Interethnic Association for the Development of the Peruvian Amazon (AIDESEP).

Source: Compiled by Worldwatch Institute from sources cited in endnote 72.

and good health care. With these established, resource management issues were approached in a way that has let villagers maintain control over their local resources by setting up community forest management committees, tree nurseries, fodder and fuel plantations, and self-help training courses for farmers and tourist-lodge owners. To aid the recuperation of depleted rhododendron forests, project staff and villagers have marked the entire conservation area off in various land use zones, including areas for strict conservation, low impact use, and intensive use. By switching from wood to kerosene for fuel, the project has eliminated the consumption of more than 1,600 kilograms of wood each day. At the initiative of villagers, subsequent programs have included family planning and literacy training for women and children. KMTNC now uses similar approaches in other Nepalese conservation areas.[74]

Halfway around the globe, in the arid mountain canyon country of western New Mexico, the Zuni Tribe has launched a comprehensive Sustainable Resource Development Plan—one of the first community-level strategies based on the Earth Summit's Agenda 21 action plan. An important outcome of this has been the Zuni Conservation Project, now staffed by up to 80 people working with the 10,000 residents of the tribe's 192,000 hectares of lands.[75]

In 1994, directed by the project's hydrologist and two technicians, a seasonal crew of 40 workers used tribal techniques to arrest soil erosion on two drainage systems covering almost 2,500 hectares. They built 300 rock and brush structures, reseeded with native grasses, closed some roads, and rebuilt older structures. The watershed restoration will help decrease siltation in the check dams (which temporarily divert water for irrigation) and in water supplies for livestock and wildlife. In connection with this restoration, other project staff work individually with ranchers to create grazing permits and management plans that reduce grazing in highly erodible or overgrazed areas, especially near streams. In a separate effort, farmers have established a seed bank to preserve customary varieties of maize, chile, squash, and other traditional crops, for both subsistence and commercial use. And the tribe has established the Nutria Conservation and Wilderness Area to preserve the rare fish that live in its waters and the numerous bird species that inhabit the canyon. "We are living proof of sustainable development," says project leader James Enote. "We have lived and prospered in the same location for thousands of years." Enote now regularly shares his experience in sustainable community development with other indigenous organizations around the world.[76]

In the mountainous region of southeastern Costa Rica and northern Panama lies the La Amistad Biosphere Reserve. Covering 1.5 million hectares, La Amistad (meaning friendship) is a complex of protected areas that includes the largest tract of tropical montane cloud forest in Central America and a critical watershed from which half of Costa Rica's freshwater originates. Within the reserve's boundaries, biologists have identified more than 10,000 species of higher plants, 400 birds, 250 reptiles and amphibians, and six tropical cats.

More than 45,000 visiting trekkers ascend the trails every year, accompanied by a similar number of porters.

In addition, La Amistad is home to four indigenous cultures—the Bribri and Cabecar of Costa Rica, and the Teribe and Guaymí of Panama—as well as to other communities that rely on the reserve's natural resources for survival. Areas around the reserve are being cleared for banana and pineapple plantations, as well as for mining, ranching, and small farming. And since the region's one million residents are already using nearly all the suitable land for farming or grazing, new immigrants are forced to carve their fields out of the remaining forest, often in the reserve's core protected area.[77]

In 1994, the McDonald's corporation, Conservation International, and Clemson University formed a partnership with a community of 80 La Amistad families. Called AMIS-CONDE (Amistad Conservation and Development Initiative), the project extends credit to participating families through a community-based bank. By selling directly to regional markets and other buyers, farmers receive higher returns for their efforts. AMISCONDE also helps identify new products, and new markets

for them. One promising candidate is naranjilla, a tart pulpy fruit
Costa Ricans commonly use in fresh drinks. Since these trees
grow on gradients as steep as 60 degrees and do best beneath a
high canopy, producing naranjillas helps to maintain critical
habitat for other plants and animals. Already, 46 farmers have
grown and sold more than 25 tons of naranjilla, for a profit of
$9,600 within a seven-month period. Naranjilla production also
employs 60 people two days a week to harvest, wash, bag, and
load the fruit onto trucks headed to market. Educational initiatives of the project include local university scholarships in forestry,
soil conservation, and other fields that directly benefit local communities. Project staff plan to expand the project to communities on the Panamanian side of the reserve.[78]

Farther south, on the Caribbean coast of northern Colombia,
a multidisciplinary group called the Fundación Pro-Sierra Nevada
de Santa Marta has assembled its resources in defense of the
world's highest coastal massif, the Sierra Nevada de Santa Marta.
Within 42 kilometers of the coast, the landscape rises to a snow-
capped peak at 5,775 meters elevation (about 18,950 feet). The
range covers about 17,000 square kilometers, roughly one-third
the area of Costa Rica. Biologically isolated from the massive
Andes to the south, the Sierra is refuge to hundreds of species
found nowhere else, including every amphibian species found
above 3,000 meters elevation. Today, the extraordinary physical and biological complexity of the Sierra is reflected in a com-
parably diverse social composite that includes 30,000 indigenous
people of the Kogi, Arhuaco, and Wiwa tribes and 150,000 peasants—not to mention three guerilla groups, several paramilitary factions, and numerous military troops. Colombia has created two national parks and two indigenous reserves in the
region. Governance of the Sierra is fragmented among three districts, 10 municipalities, and 35 government agencies, making
political coordination another fundamental challenge.[79]

Thirty-six rivers tumble down from the Sierra, providing
water to at least 1.5 million people. All of these people, in turn,
depend on the integrity of the Sierra's rich tropical forests and
the watersheds they maintain. Since the 1970s, marijuana cultivation alone has led to the deforestation of 100,000 hectares

of upland forest. Burning pasture for grazing and clearing forests for croplands have further disrupted stable water flows to the farms, villages, and cities downstream. Threats to human rights and peace are also primary issues for the region's residents, since the guerillas, drug traffickers, government, and paramilitary forces in the area engage in frequent armed conflicts.[80]

In this strife-torn region where mistrust among ethnic groups and economic classes has seemed endemic, the 90 staff members of the foundation have initiated a conservation strategy for the entire Sierra by convening a series of meetings to build consensus for needed conservation action. The foundation has encouraged a long-term and holistic approach to the process, by collecting information from each village, conducting field research, and publishing thematic maps and comprehensive studies of each major issue. Three community education centers have been established, where residents can learn both traditional and modern skills in terracing, irrigation, farming, aquaculture, literacy, and hygiene. Peasant leaders from throughout the Sierra have joined forces and formed an association of the region's ten municipalities for coordinating their conservation work. Perhaps most significantly, after eight years of careful political groundwork, the federal government recently announced that it will return 19,000 hectares of land in the Don Diego and Palomino watersheds—currently occupied by peasants and banana plantations—to their original indigenous occupants. For the Kogi, the recovery of a contiguous corridor to the coast and a full altitudinal range of ecosystems will restore a measure of integrity not just to their territory, but also to their culture.[81]

A common lesson from all of these community projects is that the great diversity of mountain ecosystems, cultures, and adaptive strategies requires a long-term commitment of personnel in order to develop effective programs. Project staff need time and training to understand local ecological systems and stresses, existing natural resource management strategies, cultural mores, community structure, and gender roles. "Sustainable mountain development is not a short-term project, it's a long-term process," says Juan Mayr-Maldonado, executive director and founder of the foundation.[82]

Integrated conservation and development projects provide no guarantee of success, however. Far from a panacea for their intended beneficiaries, these projects are best characterized as complex and time-consuming experiments in reconciling the fundamentally distinct—and often conflicting—goals of conservation and development. Many projects fail to focus on the critical linkage between these two goals. In a recent study by Stanford's Center for Conservation Biology, only 5 of 36 projects reviewed had positively contributed to the conservation of wildlife. The inescapable conclusion is that well-planned rural development does not automatically lead to the conservation of biodiversity. The authors concluded that comprehensive ecological oversight of projects is usually lacking. They recommended two types of comparative monitoring through time in unregulated areas, managed buffer zones, and core protected areas: first, assessing the total effects of the projects on biodiversity and overall ecosystem health by tracking indicator species; and second, measuring human impacts by comparing target species diversity and abundance.[83]

Since mountains are so physically, biologically, and culturally diverse, integrated and participatory programs at the community level will be imperative for achieving social and economic development that improves human lives while ensuring and even enhancing the integrity of the natural systems upon which their welfare depends. The key will be an array of local approaches, informed by lessons learned in other places and supported by the right policies at higher levels.

Moving Mountains Up the Agenda

Agendas for action in mountains need to fully represent the people who live there, because agendas made by outsiders usually fail to incorporate the local knowledge and priorities that are essential to sustainable development. A concerted effort to strengthen the voice of mountain peoples is well underway.

In preparation for the 1992 Earth Summit, a small group of mountain scientists wrote a chapter on mountains for Agenda

21, the global blueprint for sustainable development, and they persuaded the Swiss delegation to submit the chapter for ratification. The chapter recommended numerous funding priorities for sustainable mountain development. In 1994, the U.N. Commission on Sustainable Development asked governments and interested NGOs to debate and approve a plan for implementing these recommendations. The first global conference of mountain NGOs was scheduled to take place in February 1995, in Lima, Peru, to address this need. The Mountain Agenda Consultation should help coalesce an NGO mountain constituency. Already, dozens of small, effective NGOs are represented by regional associations in each major range—including the African Mountain Association, the Asia and Pacific Mountain Association, the International Commission for the Protection of the Alps, and the South American Mountain Association. If innovative partnerships are forged among mountain communities, NGOs, research institutions, and development assistance and government agencies, institutional momentum is likely to build.[84]

To elevate the status of mountain peoples and conserve their ecosystems, national governments and international development agencies will need to focus on policy reform in six areas: promoting efforts to secure land tenure or control over local resources; reducing the impacts of livestock, timber, hydropower, and minerals production in mountains; creating regional networks of conservation areas; improving knowledge about mountains through integrated research, social and environmental monitoring, and public education; establishing institutions and cooperative agreements for each major range; and integrating mountains into the projects and policies of development agencies. A key to success in each of these areas will be recognition of the unique constraints and opportunities for biological conservation and sustainable development in mountains.

Perhaps the most critical reform for deflecting growing pressures on relatively intact mountain ecosystems and cultures lies in reforming the inequitable patterns of private land ownership and public resource control that currently keep the best lands and natural resources in the hands of a small group of powerful elites. The future basis of economic security for most mountain communities lies in firm tenure or access rights to healthy ecosystems. Major components of the needed reform include comprehensive social and ecological assessments of all major development projects (a frequently violated requirement of multilateral development banks), along with genuine consultation with local people before projects affecting them are funded. Certainly, achieving this ideal will take several decades at least. Bad policies are deeply entrenched: social inequities are ancient and land has always been unequally distributed.[85]

For ethnic minorities, an achievable first step in this reform is the mapping, demarcation, and recognition of ancestral homelands, along with the creation of organizations to fight for land and resource rights through the courts. The effectiveness of this approach was demonstrated in the early 1980s, for example, by a member of Peru's Aguaruna tribe, Evaristo Nugkuag. Realizing that all Indians in the Peruvian Amazon basin were facing the same threats from cattle ranching, mining, and logging, Nugkuag helped to organize the Interethnic Association for the Development of the Peruvian Amazon (AIDESEP), a coalition of tribal groups that work collectively to protect land rights. In 1984, he went on to help found the Coordinating Body of the Federations of Amazonian Indians (COICA), representing more than 1.2 million people and 219 different tribes, many of whom live on the eastern slopes of the Andes. COICA has recently forged a strategic alliance with northern conservationists in their efforts to preserve the Amazon.[86]

In 1994, the Haisla Nation secured a commitment from the British Columbia government to protect 317,000 hectares of the Greater Kitlope Ecosystem in the province's Coast Mountains. The world's largest intact temperate rain forest, the Kitlope provides habitat for all six species of Pacific salmon and populations of North America's largest vertebrates, including black and grizzly bears, mountain goats, moose, and wolves. For several millennia, the Henaaksiala branch of the Haisla Nation has occupied the banks of the lower Kitlope River. Instead of creating a provincial park, the Haisla propose to manage the Kitlope jointly with the B.C. government, and they have created the Nanakila

Institute to develop the skills necessary for the task. This arrangement not only will allow the Haisla to continue their traditional hunting and fishing—it will also make the watershed available for scientific research and strengthen the already-successful Haisla Rediscovery program, which brings children of all ethnic backgrounds to a "rediscovery camp" to learn Haisla culture and traditions.[87]

To reduce the impacts of extractive industries in mountains, and establish a truly equitable and sustainable balance between the conservation and use of mountain resources, four systemic reforms are needed. First, and simplest, is to make far more efficient use of every tree removed and every ore extracted, and of every kilowatt of electricity and kilogram of animal product produced. If the United States, the world's largest consumer of wood, were to adopt currently available methods of waste reduction, recycling, and manufacturing efficiency, overall wood consumption could be cut at least in half. Similarly, adopting the most energy efficient technology available is usually the least expensive method for individuals, corporations, or governments to meet new demands for electricity. For reasons of immediate cost savings, resource efficiency is already revolutionizing many industries, and its logic is unassailable: by reducing waste, the same level of demand can be satisfied with sharply reduced levels of resource exploitation.[88]

Today's prices for natural resources do not even come close to telling the ecological truth.

Second is to eliminate the massive subsidies governments grant these environmentally destructive operations. Reflecting only the present costs of extraction and distribution, today's prices for natural resources do not even come close to telling the ecological truth: they ignore the full costs of denuded forests, eroded hillsides, and dammed or polluted rivers—not to mention the incalculable social costs of uprooting people living atop the resource. Recognizing full costs provides direct incentive to minimize environmental impacts, which then can yield higher returns. Farmers in Pays d'Enhaut, Switzerland refrain from using most types of fertilizers and insecticides in order to market their cheeses under a premium label created for environmentally benign products in the 1970s by the Swiss League and the local branch of the World Wildlife Fund. While prices cannot internalize the intrinsic value of a sacred forest grove or cascading river, they could substantially retard the unsustainable rates of natural resource consumption.[89]

Third is to encourage progress toward full-cost pricing by shifting a portion of taxes from income to consumption of virgin materials and resources. This will ensure that environmental costs are considered even in private consumption decisions. While this reform is unlikely to occur solely on behalf of conserving mountains, full-cost pricing is a broader policy tool for achieving sustainable development that will have substantial benefits for mountain environments—given their abundance of natural resources. Returning a portion of these taxes through microenterprise banks that invest in the communities that produced the resources could help reverse the current net transfer of wealth from mountains downhill to the plains. In addition, establishing tax credits or deductions can provide compelling incentives for land donations to conservation agencies, the sound stewardship of private lands, and other actions that benefit biodiversity.[90]

Finally, perhaps the most important task is downsizing massive extractive operations to match the small production scale best suited for fragile ecosystems and diverse cultures. For many developing countries, building new hydropower projects will probably remain a key aspect of energy strategy. Small- or medium-sized dams can be built with substantially less social and environmental impact than the massive structures currently favored by development agencies. Portions of the Andes, the Himalaya, Turkey, Ethiopia, and southwest China have numerous uninhabited sites with little or no vegetation and enormous potential for hydropower—and in many cases for small reservoirs. While even these smaller, better-sited structures would cause some damage to aquatic ecosystems, they would have far less impact than most current hydropower projects. The market already appears to grasp these fundamentals: privately financed hydro projects tend to be substantially smaller (about

100 megawatts on average), based mainly on run-of-the-river designs, and usually require little or no resettlement.[91] Counteracting mountain peoples' marginalization does not necessarily mean integrating them fully into industrial market economies—which, after all, are themselves largely unsustainable in their present form. For years, some community activists in the mountains have been advocating life-styles of simplicity instead of extravagance. Sunder Lal Bahuguna, a Gandhian leader of the Save the Himalaya movement in India, calls for using local natural resources only to achieve regional self-sufficiency, banning all future commercial logging, mining, and building of dams, and donating 12 percent of all electricity generated by large hydropower projects to local villages.[92]

Mountains comprise much of the world's remaining intact wilderness.

To shift from the goal of maximizing extraction of natural resources and commercial services will require progress towards managing resources for the health of the entire ecosystem. This shift will entail protecting viable populations of all native species, perpetuating natural-disturbance regimes on the regional scale, and allowing human uses only at levels and with methods that do not result in long-term ecological degradation. Moreover, in contrast to the quarterly intervals of stockholder reports or five-year cycles of donor project funding, conserving mountain ecosystems requires a planning timeline of centuries.[93]

To safeguard the compact variety of ecosystems found in mountains, it is crucial to create regional networks of conservation areas. Mountains comprise much of the world's remaining intact wilderness and are proportionately better protected than most other biomes, such as temperate grasslands or aquatic ecosystems. Thus, they are logical "ecological backbones" on which to build connected conservation systems. In the Andes, conservationists are developing plans for "corridors" of protected areas along the entire length of the range. A few years ago in North America, a group of conservationists dissatisfied with conventional conservation goals launched the Wildlands

Project to advocate protection and restoration of the continent's ecological integrity and diversity through the establishment of a connected system of reserves. The group proposes large, wild core reserves where biodiversity and ecological processes dominate, linked by biological corridors to allow for the natural dispersal of wide-ranging species, for genetic exchange between populations, and for migration of organisms in response to climate change. Buffers would surround the reserves and corridors and be managed to restore ecological health, nearly extinct species, and natural disturbances. In the continental United States, the two most viable prospects for regional conservation are portions of two mountain ranges: the northern Rockies and the southern Appalachians.[94]

In the Mount Everest region, an innovative set of partnerships has been created for regional conservation. A group called The Mountain Institute, based in West Virginia in the United States, has worked with the Nepalese and Chinese governments to create two multiple-use conservation areas that adjoin three existing Nepalese national parks. Together, the parks and preserves now span more than 4.1 million hectares, an area as large as Switzerland. Home to about 100,000 people, this protected area provides a full elevational range of habitats and migratory corridors for wildlife, vital to wide-ranging predators like the endangered snow leopard.[95]

Establishing regional networks of mountain protected areas will not work as an isolated reform, however. Simply creating more protected areas will not prove viable if the people who depend on these areas have no alternative means for survival. Local communities must benefit directly from protected areas before new ones can be established. Education and incentives to practice resource-conserving habits on private and community lands are critical to effective conservation. And unchecked air and water pollution, stratospheric ozone depletion, global warming, or population growth may render even the most ideal reserves unsuitable for all but the most tolerant and adaptable species. But for now, because habitat degradation and loss remain the most severe immediate threat to mountain ecosystems, protection and restoration of these ecosystems should become a high conservation priority.[96]

A fundamental impediment to raising mountains on the agendas of policymakers is lack of knowledge, combined with pervasive scientific uncertainty about the planet's most complex landscapes. Data on mountain areas—whether economic, social, or environmental— are usually incomplete when they exist at all. To date, those scant financial and technical resources that do exist have been largely spent on such narrow fields as the geomorphology of hillside crop erosion.

Local communities must benefit directly from protected areas before new ones can be established.

But more interdisciplinary research programs, along with long-term monitoring, are critical to establishing baseline scientific information on mountain ecosystems and communities, and on the rapid changes occurring there. Three examples of successful integrated research programs are the African Mountain Association's Mount Kenya Ecological Programme, the International Potato Center's Sustainable Andean Development (CONDESAN) project, and Switzerland's Man and the Biosphere project in the Alps. Mountains are also ideal settings for investigating global environmental changes, because they contain compact assemblages of diverse ecosystems that are sensitive to slight environmental changes.[97]

The dearth of widely-disseminated, accessible, and comprehensive data on mountain problems is a serious barrier to intelligent decisionmaking. The regular publication of standard indicators of the social welfare of mountain peoples and the health of their environments, such as literacy rates and forest cover, would give policymakers a much sounder basis for making decisions. These indicators could be collected by national governments, synthesized into range-by-range summaries, and periodically issued in publications such as *State of the World's Mountains: A Global Report*, which was first produced for the Earth Summit by mountain scientists, or in the quarterly journal, *Mountain Research and Development*, which has been published for the last 14 years.[98]

To assure the protection of mountain cultures and ecosystems will require forging institutional mechanisms for cooperation on transboundary problems. Movements of weather, water, soil, and animals defy political borders. The Andes and Alps, for instance, are each divided among seven countries; the Hindu Kush-Himalaya are divided among eight. Protections therefore need to be vested in cooperative agreements for each major mountain range. The Alpine Convention, negotiated in 1992 by the environment ministers of Austria, France, Germany, Italy, Liechtenstein, Slovenia, and Switzerland, provides an encouraging model. Ratified by three of these countries, the convention came into legal force in March 1995. (The European Union is also a signatory to the agreement.) The convention will ultimately include legally binding protocols to regulate Alpine tourism, transport, regional planning and sustainable development, nature protection, and mountain farming and forestry. In other international ranges, the countries involved may still be decades away from reaching comparable agreements.[99]

Eleven years ago, the eight countries that share the Hindu Kush-Himalaya ranges founded the International Centre for Integrated Mountain Development (ICIMOD) to promote sustainable development. By training scientists, development specialists, and extension workers of numerous local institutions and organizations in technical aspects of development, ICIMOD has played an important role in building human capacity in the region and raising awareness among policymakers about mountain problems. The center has coordinated research on such issues as sustainable mountain agriculture, pastoral systems, and agroforestry. Establishing similar range-wide research centers for the Andes and African mountains would provide critically needed institutional homes for integrated research and development work in those regions. Moreover, creating these centers would not require launching a new organization: they could be affiliated with an existing institution, such as the International Potato Center in Lima, Peru.[100]

Especially in industrialized countries, investment funds for integrated conservation and development projects can be raised from the private sector. An innovative model to consider is the partnership forged a few years ago between Chicago's Shorebank Corporation and Portland's Ecotrust. Shorebank is a community development bank with more than 20 years experience in increasing market opportunities and access to capital for residents of low-income neighborhoods. Ecotrust is a nonprofit organization formed to encourage investment in "conservation-based development" throughout the temperate rain forests of North America, which grow on the western slopes of the region's coastal mountains. Together, the two organizations have formed ShoreTrust, The First Environmental Bancorporation. To promote environmentally sound, community-based businesses in Ecotrust's bioregion, the two groups help to identify and create links to markets, provide technical assistance needed to expand operations, and provide access to a wide range of credit products. Shorebank is raising capital for these activities through its EcoDeposits program that features four types of bank accounts: checking, savings, money market, and certificates of deposit. Already the program has attracted over $2.6 million.[101]

Currently, not one major international institution has a distinct mandate or program for mountains.

In developing countries, genuine progress for mountain peoples will require development agencies, above all, to recognize mountain conservation and development as distinct funding priorities. Currently, not one major international institution has a distinct mandate or program for mountains: no specific departments, projects, or staff are assigned to solving development problems for mountain peoples, who are among the world's poorest people and who live in the some of the world's most complex landscapes. Analysis of 1,588 World Bank projects worth $151 billion over the last six fiscal years indicates that a mere 13, with $493 million in foreign-exchange costs, dealt directly with improving the lives of mountain peoples or protecting mountain environments. Lending records of the other multilateral development banks appear comparable. Not every institution, however, is blind to the distinct needs of mountain peoples. The Rome-based International Fund for Agricultural Development (IFAD) has focused on alleviating poverty in mountainous areas in 40 percent of its Asian projects and 70 percent of its Latin American ones.[102]

Most development investment decisions are still made by economists and engineers, for whom environmental protection and cultural conservation are at best peripheral considerations. Until ecologists and anthropologists attain equal footing and local people are included in the entire process, sustaining mountain peoples and environments will remain secondary to maximizing rates of return and erecting costly infrastructure. To fulfill their primary mandates of alleviating poverty and promoting environmentally sustainable development, the development banks will have to shift their lending priorities to strengthening the institutional capacity of the thousands of effective, grassroots groups already working throughout the world's mountains. Each year, about 4,000 northern NGOs disperse about $6 billion in development assistance, usually working with 10,000 to 20,000 southern NGOs who help up to 100 million of the world's poorest people. Ensuring that one-tenth of these funds goes directly to the tenth of humanity who live in mountains would leverage substantial new financial resources.[103]

Another potential source of funding for integrated mountain initiatives is the Global Environment Facility (GEF), the fund jointly administered by UNEP, UNDP, and the World Bank that provides grants and concessional loans to developing countries for addressing global environmental problems. Three of GEF's primary funding priorities—biodiversity, climate change, and international water issues—are particularly relevant to mountain environments. A powerful advantage of a unitary fund like GEF is that unlike traditional development lending, it is not restricted to projects designed for improvements in single sectors, such as agriculture or transportation. Funding the right sort of integrated project, then, could simultaneously satisfy more than one GEF objective. For example, reforesting the degraded slopes above a hydropower reservoir would help to conserve biodiversity, improve the quality of headwaters, and reduce the car-

bon emissions that might otherwise have resulted from deforestation.[104]

The inequitable distribution of land, natural resources,
income, and power is a major cause of both the continued
impoverishment of mountain peoples and the attendant degradation of their environment. While social inequities are ancient,
they must be reversed over the long term. Most fundamental to
this reversal will be significant reductions of materials and energy consumption by the wealthy industrialized countries, through
a use of resources that is far more efficient and far less tolerant
of excess—more consonant with real human needs. "It is industrial country growth that has to contract to free up ecological
room for the minimum growth needed in poor countries,"
writes World Bank ecologist Robert Goodland. This imperative
applies as well to the superconsumers of the developing world.
Striving for equity—between generations, regions, and nations—
provides a powerful directive for biological conservation and
sustainable development. Nothing short of a bottom-up transformation of individual values and consumer behaviors will
save the mountains and their people from the juggernaut of
development.[105]

This is a daunting challenge, and meeting it will require
decades of sustained resolve—a type of resolve more typically
associated with the world's intrepid explorers than with its con-
sumers, but now called for in all humanity. If the world's highest mountains have been able to inspire extraordinary fortitude
and ingenuity in the climbers drawn to their summits, then
the fragile ecosystems and endangered cultures from which
these pinnacles rise now merit no lesser commitment.

Notes

1. Proportion of land surface that is mountains and high plateaus from H. Louis, "Neugefasstes Höhendiagramm der Erde," *Bayer. Akad. Wiss.* (Math.- Naturwiss. Klasse, 1975), cited in Roger G. Barry, *Mountain Climate and Weather* (New York: Routledge, 1992); proportion of world population in mountains from Jack Ives, "Preface," in Peter B. Stone, ed., *State of the World's Mountains: A Global Report* (London: Zed Books Ltd., 1992); number of people dependent on mountain water from "Managing Fragile Ecosystems: Sustainable Mountain Development," Chapter 13 in United Nations, *Agenda 21: The United Nations Program of Action From Rio* (New York: U.N. Publications, 1992); Sandra Postel, *Last Oasis: Facing Water Scarcity* (New York: W.W. Norton & Company, 1992).

2. Number of people who consider a mountain sacred from Edwin Bernbaum, *Sacred Mountains of the World* (San Francisco: Sierra Club Books, 1990).

3. Map is adapted from Robert G. Bailey, *Ecosystem Geography* (New York: Springer Verlag, in press); principal map divisions are based mainly on large ecological climate zones; mountains exhibiting altitudinal zonation and having the climatic regime of adjacent lowlands are distinguished according to the character of the zonation, from Robert G. Bailey, "Ecoregions Map of the Continents" and "Explanatory Supplement to Ecoregions Map of the Continents," *Environmental Conservation*, Winter 1989; area of mountains and high plateaus is a Worldwatch Institute estimate, based on Louis, op. cit. note 1, and on earth's total land surface from Times Books, *The Times Atlas of the World* (London: Times Books, 1993); number of mountain people in 1995 should be considered a minimum and is a Worldwatch Institute estimate, based on the proportion of world population in mountains from Ives, op. cit. note 1, and on world population from Population Reference Bureau, *1994 World Population Data Sheet* (Washington, D.C.: 1994); number of countries with mountains and high plateaus is a Worldwatch Institute estimate, using criteria of at least 500 meters topographical relief or at least 1,500 meters elevation, based on *The Times Atlas of the World*, op. cit. this note, and on George Thomas Kurian, *GEO-Data: The World Geographical Encyclopedia* (Detroit, Mich.: Gale Research Company, 1989); number of ranges from Dennis G. Hanson, "Look to the Mountains," in National Geographic Society, *Mountain Worlds* (Washington, D.C.: 1988); two-fifths of mountain people in Andes, Himalaya-Hindu Kush, and African mountains is a Worldwatch Institute estimate; Table 1 based on Stone, op. cit. note 1, and on the following: Tibetan Plateau figures refer to the Chinese portion only and are from John Ackerly, director, International Campaign for Tibet, Washington, D.C., private communication, November 1, 1994; U.S. area figures from Robert G. Bailey, "Description of the Ecoregions of the United States," U.S. Forest Service (USFS), U.S. Department of Agriculture (USDA), Washington, D.C., 1994; U.S. population figures based on 1990 U.S. Census from Greg Alward, economist, Ecosystem Management, USFS, USDA, Fort Collins, Colo., unpublished printout, September 20, 1994; population of Alps from Werner Bätzing, professor of geography, Institute of Geography, University of Berne, Switzerland, private communication, August 8, 1994; Brazilian figures from José Pedro de Oliveira Costa, assessor, Andean Group, World Conservation Union (IUCN)-Brazil, São Paulo, Brazil, unpublished printout and private communication, October 6, 1994; area of Antarctica from Bailey, "Explanatory

Supplement to Ecoregions Map of the Continents," op. cit. this note; Canadian area and population from Harry Hirvonan, science advisor, State of the Environment Reporting, Environment Canada, Ottawa, Canada, private communications and unpublished printouts, September 22 and 27, 1994; the figures for Africa are highly approximate; in Africa, 45 percent of the land surface has a slope exceeding 8 degrees, more than 55 percent is above 500 meters, more than 20 percent above 1,000 meters, and about 1 percent above 2,000 meters, from Bruno Messerli et al., "African Mountains and Highlands: Introduction and Resolutions," *Mountain Research and Development*, Vol. 8, Nos. 2/3, 1988; the figure of 2 million square kilometers for the Andes should be considered a minimum; examples of mountainous regions for which area and population data are currently unavailable include: the Caribbean, Central America, Scandinavia, Greenland, the Mediterranean (e.g., Pyrenees and Apennines), non-Ukranian Carpathians, New Zealand, Australia, Yemen, Indonesia, Malaysia, the Philippines, Pacific islands, Thailand, Myanmar, Laos, and China beyond the Himalaya, Hengduan, and Tibetan Plateau; as one example of the uncounted mountain people: 400 million people live in China's mountains, according to C. Wu, "Territorial Management and Regional Development," in Geographical Society of China, ed., *Recent Development of Geographical Science in China* (Ephrata, Penn.: Science Press, 1990); population densities from "Papua New Guinea Highlands," in Stone, op. cit. note 1; Peshawar from Pitamber Sharma and Mahesh Banskota, "Population Dynamics and Sustainable Agricultural Development in Mountain Areas," in N.S. Jodha, M. Banskota, and Tej Partap, eds., *Sustainable Mountain Agriculture: Perspectives and Issues*, Vol. 1 (Kathmandu, Nepal, and New York: International Centre for Integrated Mountain Development (ICIMOD) and Intermediate Technology Publications, 1992); Francis F. Ojany, "Mount Kenya and its Environs: A Review of the Interaction between Mountains and People in an Equatorial Setting," *Mountain Research and Development*, August 1993; Elizabeth A. Byers, "Heterogeneity of Hydrologic Response in Four Mountainous Watersheds in Northwestern Rwanda," *Mountain Research and Development*, November 1991.

4. Martin Price, "The Highlands: Environmental Problems and Management Conflicts," in "Tundra and Insularity," *Biosfera* (Barcelona, Spain: *Enciclopedia Catalana*) Vol. 9, forthcoming; "Natural Hazards," in Mountain Agenda 1992, *Appeal for the Mountains* (Berne, Switzerland: Institute of Geography, University of Berne, 1992).

5. Carl Troll, "High Mountain Belts between the Polar Caps and the Equator: Their Definition and Lower Limit," *Arctic and Alpine Research*, Vol. 5, No. 3, 1973; percentage of Earth's landscape is a Worldwatch Institute estimate, based on Bailey, "Ecoregions Maps of the Continents," op. cit. note 3; Barry, op. cit. note 1; Molly Moore, "World's Wettest Place Suffers Drought-Like Conditions," *Washington Post*, October 8, 1994; Jack D. Ives et al., "The Andes: Geoecology of the Andes," in Stone, op. cit. note 1.

6. J.W. Cole, "Cultural Adaptation and Sociocultural Integration in Mountain Regions," in Social Aspects of Mountain Communities, Fourth World Congress for Rural Sociology, Torun, Poland, August 9-13, 1976; Ives et al., op. cit. note 5; Barry, op. cit. note 1; Michael A. Little, ed., "A General Prospectus on the Andean Region," *Mountain Research and Development*, February 1981.

7. Barry, op. cit. note 1; Jayanta Bandyopadhyay and Dipak Gyawali, "Himalayan Water Resources: Ecological and Political Aspects of Management," *Mountain*

Research and Development, February 1994; "African Mountain and Highland Environments: Suitability and Susceptibility," in Stone, op. cit. note 1; California from Postel, op. cit. note 1; California's population from Bureau of the Census, "Resident Population of the States," press release, U.S. Department of Commerce, Washington, D.C., July 1, 1993; reliance on mountain water is a Worldwatch Institute estimate, based on "Managing Fragile Ecosystems," op. cit. note 1, and on Ives, op. cit. note 1.

8. S.W. Buol, F.D. Hole, and R.J. McCraken, *Soil Genesis and Classification* (Ames, Iowa: Iowa State University, 1973); "Water Towers of Mankind," in Mountain Agenda 1992, op. cit. note 4; Bruno Messerli et al., eds., "Workshop on the Stability and Instability of Mountain Ecosystems," *Mountain Research and Development*, May 1983; Elizabeth Byers and Meeta Sainju, "Mountain Ecosystems and Women: Opportunities for Sustainable Development and Conservation," *Mountain Research and Development*, August 1994.

9. Byers and Sainju, op. cit. note 8; Karl S. Zimmerer, "The Loss and Maintenance of Native Crops in Mountain Agriculture," *GeoJournal*, May 1992; Jack D. Ives, "The Future of the Mountains," in Jack D. Ives, consulting ed., *Mountains: The Illustrated Library of the Earth* (Emmaus, Pa.: Rodale Press, 1994); "Biodiversity: Future Wealth," in Mountain Agenda 1992, op. cit. note 4.

10. "Biodiversity: Future Wealth," in Mountain Agenda 1992, op. cit. note 4; Jack Weatherford, *Indian Givers: How the Indians of the Americas Transformed the World* (New York: Fawcentine Columbine, 1988); Alan Thein Durning, *Guardians of the Land: Indigenous Peoples and the Health of the Earth*, Worldwatch Paper 112 (Washington, D.C.: Worldwatch Institute, December 1992).

11. Ojany, op. cit. note 3; Adam Kotarba, ed., "Special Issue on Environmental Transformation and Human Impact in the Polish Tatra Mountains," *Mountain Research and Development*, February 1992; Andes from Conservation International, *Annual Report 1993* (Washington, D.C.: 1994); Namche Barwa from Li Bosheng, professor, Institute of Botany, Chinese Academy of Sciences, Beijing, China, private communication, July 24, 1994; Sierra Nevada from Eric Beckwitt, Sierra Biodiversity Institute, North San Juan, Calif., private communication, November 3, 1994.

12. Norman Myers, "Threatened Biotas: 'Hot Spots' in Tropical Forests," *The Environmentalist*, Vol. 8, No. 3, 1988; Norman Myers, "The Biodiversity Challenge: Expanded Hot-Spots Analysis," *The Environmentalist*, Vol. 10, No. 4, 1990; Reed F. Noss and Allen Y. Cooperrider, *Saving Nature's Legacy: Protecting and Restoring Biodiversity* (Washington, D.C.: Island Press, 1994); Cynthia Carey, "Hypothesis Concerning the Causes of the Disappearance of Boreal Toads from the Mountains of Colorado," *Conservation Biology*, June 1993; World Conservation Monitoring Centre, *Global Biodiversity: Status of the Earth's Living Resources* (London: Chapman & Hall, 1992).

13. C.J. Bibby et al., *Putting Biodiversity on the Map: Priority Areas for Global Conservation* (Cambridge: International Council for Bird Preservation, 1992); number of mountainous endemic bird areas (EBAs) from geographic information system analysis of digitized information, derived from comparison of BirdLife's maps of EBAs with Bailey's map of continent's ecoregions, op. cit. note 3; analysis from Simon H. Blyth, GIS technician, Habitats Department, World Conservation Monitoring Centre, Cambridge, England, unpublished printout and map, November 1, 1994; South America from Blyth, op. cit. this note, and from Adrian J. Long, "Restricted Range and Threatened Bird Species in Tropical Montane Cloud Forests," in Lawrence S. Hamilton, James O. Juvik, and Fred N. Scatena, *Tropical Montane Cloud Forests: Proceedings of an International Symposium at San Juan, Puerto Rico, 31 May-5 June, 1993* (Honolulu, Hawaii: East-West Center Program on Environment, 1993).

14. Noss and Cooperrider, op. cit. note 12; "Biodiversity: Future Wealth," in Mountain Agenda 1992, op. cit. note 4; T.D. Schowalter, "Canopy Arthropod Community Structure and Herbivory in Old-Growth and Regenerating Forests in Western Oregon," *Canadian Journal of Forest Research*, March 1989; Sallie J. Hejl, "The Importance of Landscape Patterns to Bird Diversity: A Perspective from the Northern Rocky Mountains," *Northwest Environmental Journal*, Vol. 8, No. 1, 1992; range of tropical mountain ecosystems from Lawrence S. Hamilton, "Mountain Chronicle: Status and Current Developments in Mountain Protected Areas," *Mountain Research and Development*, August 1993; Samuel M. Scheiner and José M. Rey-Benayas, "Global Patterns of Plant Diversity," *Evolutionary Ecology*, Vol. 8, No. 4, 1994.

15. "Cultural Diversity," in Mountain Agenda 1992, op. cit. note 4; Byers and Sainju, op. cit. note 8; Elizabeth May, "Women: The Resource Managers," *Our Planet*, Vol. 6, No. 4, 1994; Suresh Chand Rai, Eklabya Sharma, and Rakesh Chandra Sundriyal, "Conservation in the Sikkim Himalaya: Traditional Knowledge and Land-Use of the Mamlay Watershed," *Environmental Conservation*, Spring 1994.

16. Agarwal quotation from Centre for Science and Environment, *State of India's Environment, A Citizen's Report: Floods, Flood Plains, and Environmental Myths* (New Delhi: 1991); Durning, op. cit. note 10.

17. Throughout this paper, "indigenous," "native," and "tribal" are terms used interchangeably to describe the people and cultures that originally inhabited the land before colonization or conversion occurred; for a discussion of the various definitions and numbers of "indigenous peoples," and for the number of Quechua, see Durning, op. cit. note 10; Art Davidson, *Endangered Peoples* (San Francisco: Sierra Club Books, 1993); Robert W. Kates, B.L. Turner, and William C. Clark, "The Great Transformation," in B.L. Turner et al., eds., *The Earth as Transformed by Human Action: Global and Regional Changes in the Biosphere over the Last 300 Years* (New York: Cambridge University Press, 1990); Badenkov et al., "Mountains of the Former Soviet Union: Value, Diversity, Uncertainty," in Stone, op. cit. note 1.

18. Rural Andeans from Hugo Li Pun, International Development Research Centre, "Sustainable Andean Development: Project Summary," Ottawa, Canada, 1992, and from George Psacharopoulos and Harry A. Patrinos, "Indigenous People and Poverty In Latin America," *Finance and Development*, March 1994; number of Chinese from World Bank, *China: Poverty Reduction Strategies for the 1990s* (Washington, D.C.: 1992); majority of poor Chinese being ethnic minorities in mountains from Jack D. Ives, *Children and Poverty in Mountains* (New York: UNICEF Environment Section, in press); Appalachian poverty from D. Schnelling et al., "The Appalachians of North America: Marginal in the Midst of Plenty," in Stone, op. cit. note 1.

19. Jack D. Ives, "Mountain Environments," in G.B. Marini-Bettolo, *A Modern Approach to the Protection of the Environment* (Rome: Pontificiae Academiae Scientiarvm, 1989); Nigel J.R. Allan, "Introduction," in Nigel J.R. Allan, Gregory W. Knapp, and Christoph Stadel, eds., *Human Impact on Mountains* (Lanham, Md.: Rowman & Littlefield Publishers, 1988); "bhotias" from Jayanta Bandyopadhyay, director of research, International Academy for the Environment, Geneva, Switzerland, private communication, October 6, 1994; Amnesty International USA, *Amnesty International Report 1994* (New York: 1994); U.S. Department of State, *Country Reports on Human Rights Practices for 1993*, submitted to the Committee on Foreign Affairs, U.S. House of Representatives, and the Committee on Foreign Relations, U.S. Senate, February 1994.

20. "From Marginality to Development," in Mountain Agenda 1992, op. cit. note 4.

21. Major armed conflicts (involving more than 1,000 deaths) from Stockholm International Peace Research Institute (SIPRI), *SIPRI Yearbook 1994* (New York: Oxford University Press, 1994), with number based in mountains being a Worldwatch Institute estimate; list of mountain areas damaged by military action is indicative, not inclusive; Ives, "The Future of the Mountains," op. cit. note 9.

22. Byers and Sainju, op. cit. note 8; H. Byron Earhart, "Sacred Mountains in Japan: Shugendo as 'Mountain Religion'," in Michael Charles Tobias and Harold Drasdo, eds., *The Mountain Spirit* (Woodstock, N.Y.: Overlook Press, 1979); examples of sacred mountains from Bernbaum, op. cit. note 2.

23. A. John De Boer, "Sustainable Approaches to Hillside Agricultural Development," in H. Jeffrey Leonard et al., *Environment and the Poor: Development Strategies for a Common Agenda* (New Brunswick, N.J.: Transaction Books, for Overseas Development Council, 1989); N.S. Jodha, M. Banskota, and Tej Partap, "Strategies for the Sustainable Development of Mountain Agriculture: An Overview," in Jodha, Banskota, and Partap, op. cit. note 3; Alan B. Durning, *Poverty and the Environment: Reversing the Downward Spiral*, Worldwatch Paper 92 (Washington, D.C.: Worldwatch Institute, November 1989); Bryan Carson, *The Land, the Farmer, and the Future: A Soil Fertility Management Strategy for Nepal*, ICIMOD Occasional Paper No. 25 (Kathmandu, Nepal: ICIMOD, 1992); Madhav Gadgil, "Biodiversity and India's Degraded Lands," *Ambio*, May 1993.

24. Jodha, Banskota, and Partap, op. cit. note 23; Byers and Sainju, op. cit. note 8.

25. Ives et al., op. cit. note 5.

26. Jodha, Banskota, and Partap, op. cit. note 23; "Crop Species Disappearing in Garhwal," *Down to Earth*, August 15, 1992; Karl S. Zimmerer, "The Loss and Maintenance of Native Crops in Mountain Agriculture," *GeoJournal*, May 1992.

27. Reed F. Noss, "Cows and Conservation Biology," *Conservation Biology*, September 1994; Thomas L. Fleischner, "Ecological Costs of Livestock Grazing in Western North America," *Conservation Biology*, September 1994; Alan B. Durning and Holly B. Brough, *Taking Stock: Animal Farming and the Environment*,

Worldwatch Paper 103 (Washington, D.C.: Worldwatch Institute, July 1991).

28. Leonard Berry, Laurence A. Lewis, and Cara Williams, "East African Highlands," in Turner et al., op. cit. note 17; Ives et al., op. cit. note 5; Rwandan arable land from U.N. Food and Agriculture Organization (FAO), *Production Yearbook 1993* (Rome: 1994); Byers, op. cit. note 3; Nepal from Jodha, Banskota, and Partap, op. cit. note 23.

29. Byers and Sainju, op. cit. note 8; Ilse Marks, "Andean Women and Food Technology Contest," *Cooperation South*, September 1994; Nepalese survey from Centre for Women and Development (CWD), *Women's Work and Family Strategies Under Conditions of Agricultural Modernization* (Kathmandu, Nepal: 1988), cited in Byers and Sainju, op. cit. note 8; Meena Archarya and L. Bennett, *Women and the Subsistence Sector: Economic Participation and Household Decision-Making in Nepal*, World Bank Staff Working Paper 526 (Washington, D.C.: World Bank, 1983).

30. William C. Thiesenhusen, ed., *Searching for Agrarian Reform in Latin America* (Boston: Unwin Hyman, 1989), cited in Nancy R. Forester, "Protecting Fragile Lands: New Reasons to Tackle Old Problems," *World Development*, Vol. 20, No. 4; Ed Riddell, Vecinos Mundiales, Santiago, Chile, private communication, July 21, 1994.

31. Kenneth Hewitt, "Mountain Hazards," *GeoJournal*, May 1992; Bertil Lintner, "Opium War," *Far Eastern Economic Review*, January 20, 1994; Victor Mallett, "'Golden Quadrangle' United by a Desire to Make Money," *Financial Times*, March 3, 1994; James McGregor, "The Opium War: Burma Road Heroin Breeds Addicts, AIDS along China's Border," *Wall Street Journal*, September 29, 1992; "Drug War High in the Andes," *The Economist*, February 13, 1993; Peru coca deforestation from Marc Dourojeanni, senior environmental adviser, World Bank, Washington, D.C., private communication, October 1989, cited in "Latin America: Resource Environment Overview," in World Resources Institute, *World Resources 1990/91* (New York: Oxford University Press, 1990).

32. Werner Bätzing, "The Alps: An Ecosystem in Transformation," in Stone, op. cit. note 1; Price, op. cit. note 4.

33. Hans Hurni et al., "African Mountain and Highland Environments: Suitability and Susceptibility," in Stone, op. cit. note 1.

34. Erik P. Eckholm, *Losing Ground: Environmental Stress and World Food Prospects* (New York: W.W. Norton & Company, 1976); Bryan Carson, *Erosion and Sedimentation Processes in the Nepalese Himalaya*, Occasional Paper No. 1 (Kathmandu, Nepal: ICIMOD, 1985); Lawrence S. Hamilton, "What Are the Impacts of Himalayan Deforestation on the Ganges-Brahmaputra Lowlands and Delta? Assumptions and Facts," *Mountain Research and Development*, August 1987; D.A. Gilmour, "Not Seeing the Trees for the Forest: A Re-Appraisal of the Deforestation Crisis in Two Hill Districts of Nepal," *Mountain Research and Development*, November 1988; Jack D. Ives and Bruno Messerli, *The Himalayan Dilemma: Reconciling Conservation and Development* (New York: Routledge and the United Nations University, 1989); J.S. Rawat and M.S. Rawat, "The Nana Kosi Watershed, Central Himalaya, India. Part II: Human Impacts on Stream Runoff," *Mountain Research and Development*, August 1994; Byers and Sainju, op. cit. note 8.

35. Ramachandra Guha, *The Unquiet Woods: Ecological Change and Peasant Resistance in the Himalaya* (Berkeley: University of California Press, 1990).

36. Byers and Sainju, op. cit. note 8.

37. FAO, *Forest Resources Assessment 1990: Tropical Countries*, FAO Forestry Paper 112 (Rome: 1993); Guatemala and Bolivia from Byers and Sainju, op. cit. note 8; "Forces on the Forests," in Mountain Agenda 1992, op. cit. note 4; Lawrence S. Hamilton, "The Protective Role of Mountain Forests," *GeoJournal*, May 1992.

38. Lawrence S. Hamilton, James O. Juvik, and Fred N. Scatena, "The Puerto Rico Tropical Cloud Forest Symposium: Introduction and Workshop Synthesis," in Hamilton, Juvik, and Scatena, op. cit. note 13; Lawrence S. Hamilton, vice chair for mountains, Commission on National Parks and Protected Areas (CNPPA), "A Campaign for Cloud Forests: Unique and Valuable Ecosystems At Risk," IUCN Focus Series, IUCN, Gland, Switzerland, December 1994; proportion of restricted-range birds from Long, op. cit. note 13.

39. Hamilton, Juvik, and Scatena, op. cit. note 38.

40. Erin Kellogg, ed., "Coastal Temperate Rain Forests: Ecological Characteristics, Status, and Distribution Worldwide" (a working manuscript), Occasional Paper Series No. 1, Ecotrust/Conservation International, Portland, Ore., June 1992; Spencer B. Beebe and Edward C. Wolf, "The Coastal Temperate Rain Forest: An Ecosystem Management Approach," in Keith Moore, *An Inventory of Watersheds in the Coastal Temperate Forests of British Columbia* (Queen Charlotte City, B.C., and Portland, Ore.: Earthlife Canada Foundation and Ecotrust/Conservation International, 1991); Derek Denniston, "Conserving the Other Rain Forest," in Lester R. Brown, Hal Kane, and David Malin Roodman, *Vital Signs 1994* (New York: W.W. Norton & Company, 1994).

41. Noss and Cooperrider, op. cit. note 12; Bill Devall, ed., *Clearcut: The Tragedy of Industrial Forestry* (San Francisco: Sierra Club Books/Earth Island Press, 1993); Beckwitt, op. cit. note 11; John Hof, U.S. Forest Service, U.S. Department of Agriculture, Fort Collins, Colo., private communication, December 6, 1994.

42. David J. Fox, "Mining and Damming," in Ives, *Mountains*, op. cit. note 9.

43. Robert Goodland, "Environmental Sustainability in the Power Sector," *Impact Assessment*, Vol. 12, No. 4, 1994 and Vol. 13, No. 1, 1995; Malaysia from Kieran Cooke, "Domestic Funding for Borneo Dam," *Financial Times*, December 12, 1994.

44. Three Gorges from Goodland, op. cit. note 43, and from Patricia Adams, "Planning for Disaster: China's Three Gorges Dam," *Multinational Monitor*, September 1993; N.D. Jayal, "Tehri Clearance Fraught with Grave Consequences," *Amrita Bazar Patrika*, March 17, 1994; Bandyopadhyay and Gyawali, op. cit. note 7.

45. Elliott Mainzer, "Moving Lesotho's Water: Is South Africa's Gain Lesotho's Loss?," *World Rivers Review*, Fourth Quarter 1994.

46. International Rivers Network, "Manibelli Declaration," Berkeley, Calif., October 1994; Derek Denniston, "A Moment of Truth," *World Watch*, January/February 1995; Juliette Majot, "Resistance Grows as Plans Move Ahead to Dam the Bio Bio," *World Rivers Review*, Fourth Quarter 1993.

47. John E. Young, *Mining the Earth,* Worldwatch Paper 109 (Washington, D.C.: Worldwatch Institute, July 1992); Thomas J. Hilliard, "Stibnite—The Next Summitville?," *Clementine,* Autumn 1993; EPA cleanup estimate from Rick Young and Dan Noyes, "The Road to Summitville, Gold Mining Debacle," *New York Times,* August 14, 1994; Peter Nielsen and Bruce Farling, "Hazardous Wastes Endanger Water, Wildlife, Land: Mining Catastrophe in Clark Fork," *Clementine,* Autumn 1991.

48. James Brooke, "For U.S. Miners, the Rush Is on to Latin America," *New York Times,* April 17, 1994; Abhinandan Bhardwaj and Sunil Dhar, Centre for Advanced Study of Geology, Panjab University, Chandigarh, India, private communication, September 2, 1993.

49. Minewatch, "Ok Tedi," Minewatch Briefing No. 13, London, June 1991; "BHP Faces Huge Damages Claim," *Mining Journal,* May 6, 1994; John Gordon, Slater & Gordon, Melbourne, Australia, background paper, December 1994.

50. Lester R. Brown, "World Economy Expanding," in Brown, Kane, and Roodman, op. cit. note 43; Alan Thein Durning, *How Much is Enough? The Consumer Society and the Future of the Earth* (New York: W.W. Norton & Company, 1992).

51. John Muir, *John of the Mountains* (Madison: University of Wisconsin Press, 1975).

52. Erwin F. Groetzbach, "High Mountains as Human Habitat," in Allan, Knapp, and Stadel, op. cit. note 19; Figure 1 from Kenneth E. Hornback, chief, Socio-Economic Studies, National Park Service, Fort Collins, Colo., unpublished print-out and private communication, August 26, 1994; John Mitchell, "Our National Parks: Legacy at Risk," *National Geographic,* October 1994; Yu-fai Leung and Jeffrey L. Marion, "Trail Degradation as Influenced by Environmental Factors: A State of the Knowledge Review," *Journal of Soil and Water Conservation,* forthcoming.

53. Kim Larsen, ed., "Abode of Snows," *Buzzworm,* March/April 1993; N.K. Doval, "Leave the Fragile Hills Alone," *Hindu Magazine,* November 6, 1994; Gangotri from Edwin Bernbaum, Commission on National Parks and Protected Areas (CNPPA), "Recommendations for Using Cultural Resources to Enhance the Gangotri Conservation Project in the Indian Himalaya," IUCN, Gland, Switzerland, September 1994; Byers and Sainju, op. cit. note 8; Nepal from Cherry Enoki, "Tourism by the Numbers," *World Paper,* September 1994.

54. International Commission for the Protection of the Alps (CIPRA), "An Alpine Convention—A Solution for an Avalanche of Problems," press release, Munich, Germany, September 1994; number of jobs from Christian Pfister and Paul Messerli, "Switzerland," in Turner et al., op. cit. note 17; Derek Denniston, "Alpine Slide," *World Watch,* September/October 1992; Werner Bätzing, Manfred Perlik, and Majda Dekleva, "Urbanization and Depopulation in the Alps: An Analysis of Current Socioeconomic Structural Change Based on Communal and Regional Development Types," *Mountain Research and Development,* forthcoming.

55. Alpine golf courses from "Seven-Nation Convention to Protect Alps Expected to Come into Force by End of Year," *International Environment Reporter,* September

21, 1994; Anne E. Platt, "The Trouble with Golf," *World Watch*, May/June 1994; Malaysia from Larry Hamilton, vice chair for mountains, Commission on National Parks and Protected Areas (CNPPA), IUCN, Hinesburg, Vt., private communication, July 23, 1994; China from Jack Ives, professor of geoecology, Division of Environmental Studies, College of Agricultural and Environmental Sciences, University of California at Davis, private communication, September 28, 1994; Philip Shenon, "FORE! Golf in Asia Hits Environmental Rough," *New York Times*, October 22, 1994.

56. "Alpine Skiing: The Dangers of Overdevelopment," in Ives, *Mountains*, op. cit. note 9; Jack D. Ives, "Mountains North and South," in Stone, op. cit. note 1; Shawn Emery, "Helicology," *Summit*, Spring 1994; D.J. Herda and L.A. Pifer, "Fast and Loose on the Slopes," *Washington Post*, November 14, 1993; Dirk Johnson, "The Battle Over Man-Made Snow: Environmentalists Confront Ski Resorts on Diverting Water," *New York Times*, November 14, 1994; Shiro Tsuyuzaki, "Environmental Deterioration Resulting from Ski Resort Construction in Japan," *Environmental Conservation*, Summer 1994.

57. Alex Markels, "Moving to the Mountains," *Snow Country*, July/August 1994; population of Aspen based on 1991 town census, from Kathleen Strickland, chief deputy clerk, Town of Aspen, Colo., private communication, December 9, 1994; Kenneth Labich, "The Geography of an Emerging America," *Fortune*, June 27, 1994.

58. Labich, op. cit. note 57; Jordan Bonfante, "Sky's the Limit," *Time*, September 6, 1993; Ray Rasker and Dennis Glick, "Footloose Entrepreneurs: Pioneers of the New West?" *illahee* (Institute of Environmental Studies, University of Washington, Seattle), Spring 1994; Markels, op. cit. note 57.

59. Ray Rasker, Norma Tirrell, and Deanne Kloepfer, *The Wealth of Nature: New Economic Realities in the Yellowstone Region* (Washington, D.C.: The Wilderness Society, 1992).

60. Increases since 1950 from Pfister and Messerli, op. cit. note 54; "Seven-Nation Convention to Protect Alps," op. cit. note 55; Alan Riding, "Swiss Give New Meaning to Roadblock," *New York Times*, February 28, 1994; Kotarba, op. cit. note 11; Derek Denniston, "Air Pollution Damaging Forests," in Lester R. Brown, Hal Kane, and Ed Ayres, *Vital Signs 1993: The Trends That Are Shaping Our Future* (New York: W.W. Norton & Company, 1993); Christine L. Shaver, Kathy A. Tonnessen, and Tonnie G. Maniero, "Clearing the Air at Great Smoky Mountains National Park," *Ecological Applications*, Vol. 4, No. 4, 1994.

61. Intergovernmental Panel on Climate Change (IPCC), *Climate Change 1992: The IPCC Supplementary Report* (New York: Cambridge University Press, 1992); Derek Denniston, "Icy Indicators of Warming," *World Watch*, January/February 1993; Johannes Oerlemans, "Quantifying Global Warming from the Retreat of Glaciers," *Science*, April 8, 1994.

62. Robert L. Peters and Thomas E. Lovejoy, eds., *Global Warming and Biological Diversity* (New Haven, Conn.: Yale University Press, 1992); IPCC, "Draft Summary for Policymakers of the 1994 Working Group I Report on Radiative Forcing of Climate Change," Maastricht, The Netherlands, September 15, 1994; Noss and Cooperrider, op. cit. note 12.

63. Georg Grabherr, Michael Gottfried, and Harald Pauli, "Climate Effects on Mountain Plants," *Nature*, June 9, 1994.

64. Hamilton, Juvik, and Scatena, op. cit. note 38; Zhang Xinshi, "A Vegetation-Climate Classification System for Global Change Studies in China," *Quaternary Sciences*, No. 2, 1993; Robert L. Peters, "Effects of Global Warming on Species and Habitats: An Overview," *Endangered Species Update*, May 1988; Noss and Cooperrider, op. cit. note 12.

65. Patrick N. Halpin, "GIS Analysis of the Potential Impacts of Climate Change on Mountain Ecosystems and Protected Areas," in Martin F. Price and D. Ian Heywood, eds., *Mountain Environments & Geographic Information Systems* (London: Taylor & Francis, 1994); Postel, op. cit. note 1.

66. Tongariro Mountains from P.H.C. Lucas, "History and Rationale for Mountain Parks as Exemplified by Four Mountain Areas of Aotearoa (New Zealand)," in Lawrence S. Hamilton, Daniel P. Bauer, and Helen F. Takeuchi, eds., *Parks, Peaks, and People: A Collection of Papers Arising from an International Consultation on Protected Areas in Mountain Environments, Held In Hawaii Volcanoes National Park, October 26-November 2, 1991* (Honolulu, Hawaii: East-West Center Program on Environment, 1993); Hamilton, op. cit. note 14; 8 percent figure is a Worldwatch Institute estimate, based on total area of mountain protected areas from James W. Thorsell and Jeremy Harrison, "National Parks and Nature Reserves in Mountain Environments of the World," *GeoJournal*, May 1992, and on total area of world's mountains, op. cit. note 3; Table 5 includes protected areas over 10,000 hectares in size with a minimum of 1,500 meters of relative relief in IUCN's management classes I-IV (Strict Nature Preserve, National Park, National Monument/Natural Landmark, and Wildlife Sanctuary); if the minimum relief had been lowered to 1,000 meters, the number of protected areas would have increased by 200; production-oriented areas, such as timber reserves or national forests, are excluded.

67. Noss and Cooperrider, op. cit. note 12; Thorsell and Harrison, op. cit. note 66; Duncan Poore, ed., *Guidelines for Mountain Protected Areas*, Commission on National Parks and Protected Areas (CNPPA), IUCN Protected Area Programme Series No. 2 (Gland, Switzerland: IUCN, 1992); Mark Clayton, "Bears in the Canadian Rockies May Be Fighting a Losing Battle," *Christian Science Monitor*, November 8, 1994.

68. Noss and Cooperrider, op. cit. note 12; A.W. Gentry, "Endemism in Tropical Versus Temperate Plant Communities," in Michael E. Soulé, ed., *Conservation Biology: The Science of Scarcity and Diversity* (Sunderland, Mass.: Sinauer Associates, 1986); Shahid Naeem et al., "Declining Biodiversity Can Alter the Performance of Ecosystems," *Nature*, April 21, 1994; David Tilman and John A. Downing, "Biodiversity and Stability in Grasslands," *Nature*, January 27, 1994.

69. U.S. Congress, Office of Technology Assessment, *Harmful Non-Indigenous Species in the United States* (Washington, D.C.: U.S. Government Printing Office, 1993); Kevin O'Connor, "'Invaders': Plant and Animal Invasions of Mountain Ecosystems and Implications for Protected Area Management," in Hamilton, Bauer, and Takeuchi, op. cit. note 66; Hamilton, Juvik, and Scatena, op. cit. note 38; Mark Cheater, "Alien Invasion," *Nature Conservancy*, September/October 1992.

70. UNESCO, *Task Force on Criteria and Guidelines for the Choice and Establishment of Biosphere Reserves*, Man and the Biosphere Report No. 22 (Paris: 1974); number of mountain biosphere reserves should be considered an estimate and is based on the number with a relative relief exceeding 1,500 meters, from Thomas Schaaf, programme specialist, Division of Ecological Sciences, United Nations Educational, Scientific, and Cultural Organization (UNESCO), Paris, France, private communication, June 15, 1994; R.F. Dasmann, "Biosphere Reserves, Buffers, and Boundaries," *BioScience*, Vol. 38, 1988; M.I. Dyer and M.M. Holland, "The Biosphere-Reserve Concept: Needs for a Network Design," *BioScience*, May 1991; Noss and Cooperrider, op. cit. note 12.

71. Aldo Leopold, *A Sand County Almanac* (New York: Oxford University Press, 1949); Paul G. Sneed, "Learning to Think Like a Mountain: A Review of Cooperative Management Regimes Appropriate for Mountain Protected Areas," in Hamilton, Bauer, and Takeuchi, op. cit. note 66.

72. Michael Wells and Katrina Brandon, *People and Parks: Linking Protected Area Management with Local Communities* (Washington, D.C.: World Bank, World Wildlife Fund, and U.S. Agency for International Development, 1992); James E. Enote and Alisa A. Mallari, "Indigenous Cultures and Sustainable Development: Two Years After the UNCED Conference," Working Group Paper Indigenous Issues, 1994; Byers and Sainju, op. cit. note 8; Table 6 from The Mountain Institute, *Nepal's Newest National Park: Makalu-Barun National Park and Conservation Area, Annual Report 1992* (Franklin, W. Va.: 1993); Task Force, *Makalu-Barun Conservation Project Management Plan* (Kathmandu, Nepal, and Franklin, W. Va.: 1991); Alton Byers, senior conservationist and protected area specialist, The Mountain Institute, Franklin, W. Va., private communication, October 17, 1994; "Tuenjai Deetes, Asia Winner," in "Detailed Background Information on the 1994 Fifth Annual Goldman Environmental Prize Winners," Goldman Environmental Foundation, San Francisco, Calif., April 18, 1994; Mattole from Freeman House, "To Learn the Things We Know," in Richard Nilsen, ed., *Helping Nature Heal: An Introduction to Environmental Restoration* (Berkeley, Calif.: Whole Earth Catalog/Ten Speed Press Books, 1991), and from Peter Berg, "Putting 'Bio' in front of 'Regional'," *Landscape Architecture*, April 1994; World Neighbors, *Beyond Cairo: The Integration of Population and Environment in Baudha-Bahunipati, Nepal* (Oklahoma City, Okla.: 1994); AIDESEP from Bruce Cabarle, Manuel Huaya Panduro, and Oswaldo Manihuari Murayari, "Ecofarming in the Peruvian Amazon: the Integrated Family and Communal Gardening Project (HIFCO)," prepared for the Liz Claiborne and Art Ortenberg Foundation Community Based Conservation Workshop, Airlie, Va., October 18-22, 1994.

73. John L. Hough and Mingma Norbu Sherpa, "Bottom Up vs. Basic Needs: Integrating Conservation and Development in the Annapurna and Michiru Mountain Conservation Areas of Nepal and Malawi," *Ambio*, Vol. 18, No. 8, 1989; Michael P. Wells, "A Profile and Interim Assessment of the Annapurna Conservation Area Project, Nepal," prepared for the Claiborne-Ortenberg Foundation Workshop, op. cit. note 72; Chandra P. Gurung, "Linking Biodiversity Conservation to Community Development: Annapurna Conservation Area Project Approach to Protected Area Management," paper presented at *The Regional Seminar on Community Development and Conservation of Biodiversity through Community Forestry*, Regional Community Forestry Training Centre (RECOFTC),

Kasetsart University, Bangkok, Thailand, October 26-28, 1994.

74. Wells and Brandon, op. cit. note 72; Hough and Sherpa, op. cit. note 73; "The Annapurna Conservation Area Project, Nepal: A Case Study," Annex B, in Michael Brown and Barbara Wyckoff-Baird, *Designing Integrated Conservation and Development Projects* (Washington, D.C.: Biodiversity Support Program, 1992); Gurung, op. cit. note 73.

75. Pueblo of Zuni, Zuni Conservation Project, "Zuni Sustainable Resource Development," Zuni, N.M., 1993; James Enote, project leader, Pueblo of Zuni, Zuni Conservation Project, Zuni, N.M., private communications, July 25, 1994 and December 16, 1994.

76. Stanford Lalio, geographic information systems coordinator, Pueblo of Zuni, Zuni Conservation Project, Zuni, N.M., private communication, October 14, 1994; Pueblo of Zuni, op. cit. note 75; Enote, op. cit. note 75; James Enote, "Saving the Land and Preserving the Culture: Environmentalism at the Pueblo Zuni," in Barbara R. Johnson, ed., *Who Pays the Price? Examining the Sociocultural Context of Environmental Crisis* (Oklahoma City, Okla.: Society for Applied Anthropology, 1993).

77. Conservation International, "Clemson University, Conservation International, and McDonald's Join Forces in New Approach to Sustainable Development," press release, August 17, 1994; Conservation International, op. cit. note 11.

78. Ibid.

79. Juan Mayr Maldonado, executive director, Fundación Pro-Sierra Nevada de Santa Marta, Bogotá, Colombia, private communication at Worldwatch Institute, Washington, D.C., November 1, 1994; Fundación Pro-Sierra Nevada de Santa Marta, "Proposal Presented to John D. and Catherine T. MacArthur Foundation," Santafé de Bogotá, April 1993; Elizabeth Kemf, "Aluna: The Place Where the Mother Was Born," in Elizabeth Kemf, ed., *The Law of the Mother* (San Francisco: Sierra Club Books, 1993).

80. Mayr, op. cit. note 79; Fundación Pro-Sierra Nevada de Santa Marta, op. cit. note 79.

81. Ibid.

82. Liz Claiborne and Art Ortenberg Foundation, *The View from Airlie: Community Based Conservation in Perspective*, prepared for the Claiborne-Ortenberg Foundation Workshop, op. cit. note 72; David Western and Michael Wright, "Issues in Community Based Conservation," prepared for the Claiborne-Ortenberg Foundation Workshop, op. cit. note 72; Mayr quotation from Mayr, op. cit. note 79.

83. Wells and Brandon, op. cit. note 72; Claire Kremen, Adina M. Merenlender, and Dennis Murphy, "Ecological Monitoring: A Vital Need for Integrated Conservation and Development Programs in the Tropics," *Conservation Biology*, June 1994.

84. Derek Denniston for The Mountain Institute, *Report on the NGO Workshop on the Mountain Agenda*, Spruce Knob, W. Va., July 22-26, 1994 (Franklin, W. Va.: The Mountain Institute, 1994); Derek Denniston, "Making Room for Mountains," *World Watch*, November/December 1993; Jack D. Ives, "Editorial," *Mountain*

Research and Development, August 1994; South American Mountain Association will formally replace the Andean Mountain Association at a conference scheduled for April, 1995 in La Paz, Bolivia, from Ives, op. cit. note 55.

85. John C. Ryan, *Life Support: Conserving Biological Diversity*, Worldwatch Paper 108 (Washington, D.C.: Worldwatch Institute, April 1992); Durning, op. cit. note 10.

86. Derek Denniston, "Defending the Land with Maps," *World Watch*, January/February 1994; Durning, op. cit. note 10; Goldman Environmental Foundation, "1991 Goldman Environmental Prize Winners," press release, San Francisco, Calif., April 22, 1991.

87. Ian Gill, "An Easy Million," *Georgia Straight*, August 19-26, 1994; Bart Robinson, "Earthly Treasure," *Equinox*, September/October 1994; Ian Gill, director, Ecotrust Canada, Vancouver, Canada, private communication, January 3, 1995.

88. Figure of at least one-half reduction does not include significant reductions available through more efficient building construction, from John C. Ryan, director of research, Northwest Environment Watch, Seattle, Wash., private communication, October 27, 1994, and from Table 5-4 in Sandra Postel and John C. Ryan, "Reforming Forestry," in Lester R. Brown et al., *State of the World 1991* (New York: W.W. Norton & Company, 1991); John E. Young and Aaron Sachs, *The Next Efficiency Revolution: Creating a Sustainable Materials Economy*, Worldwatch Paper 121 (Washington, D.C.: Worldwatch Institute, September 1994); Christopher Flavin and Nicholas Lenssen, *Powering the Future: Blueprint for a Sustainable Electricity Industry*, Worldwatch Paper 119 (Washington, D.C.: Worldwatch Institute, June 1994).

89. Young, op. cit. note 47; Alan Thein Durning, *Saving the Forests: What Will It Take?*, Worldwatch Paper 117 (Washington, D.C.: Worldwatch Institute, December 1993); Robert Hutchison, "The Alpine Farmer as Conservationist," *Ceres*, November-December 1993.

90. Sandra Postel and Christopher Flavin, "Reshaping the Global Economy," in Brown et al., op. cit. note 88; Derek Denniston, "Saving the Himalaya," *World Watch*, November/December 1993; Noss and Cooperrider, op. cit. note 12.

91. Goodland, op. cit. note 43; Robert Goodland, "Ethical Priorities in Environmentally Sustainable Energy Systems: The Case of Tropical Hydropower," *International Journal of Sustainable Development*, Vol. 1, No. 4; privately financed projects from Christopher Flavin, vice president for research, Worldwatch Institute, Washington, D.C., private communication about staff panel on hydropower held at the World Bank, Washington, D.C., November 16, 1994.

92. Sunder Lal Bahuguna, "A Practical Step Towards a Sustainable Society: Development Re-Defined," Save Himalaya Movement, Ganga-Himalaya Kuti, 1994; Sunder Lal Bahuguna, "Strategies for Sound Development in the Himalaya," in Tej Vir Singh and Jagdish Kaur, eds., *Studies in Himalayan Ecology and Development Strategies* (New Delhi: Himalayan Books, 1989).

93. Erin Kellogg and Kirk Johnson, *Double Green Policy: Towards Environmental and Economic Prosperity, Middle Fork Conference '93* (Portland, Ore., and Seattle, Wash.: Ecotrust and Northwest Policy Center, 1993); R. Edward Grumbine, "What is

Ecosystem Management?" *Conservation Biology*, March 1994; Noss and Cooperrider, op. cit. note 12.

94. Reed Noss and Larry Harris, "Nodes, Networks, and MUMs: Preserving Diversity at All Scales," *Environmental Management*, Vol. 10, 1986; Larry Hamilton, vice chair for mountains, Commission on National Parks and Protected Areas (CNPPA), IUCN, "Mountain Protected Area Update," Islands and Highlands, Hinesburg, Vt., September 28, 1994; Dave Foreman et al., "The Wildands Project," *Earth Island Journal*, Spring 1993; David Johns, Wildlands Project, McMinnville, Ore., private communication, December 8, 1994; Noss and Cooperrider, op. cit. note 12.

95. Daniel Taylor-Ide, Alton C. Byers, and Gabriel J. Campbell, "Mountains, Nations, Parks, and Conservation," *GeoJournal*, May 1992; area of reserves from Nepal National Conservation Strategy Implementation Programme, National Planning Commission, His Majesty's Government of Nepal and IUCN, *Background Papers to the National Conservation Strategy for Nepal, Vol. II* (Kathmandu, Nepal, and Gland, Switzerland: 1991); Mount Everest Ecosystem Program, *Qomolangma Nature Preserve Project, Annual Report 1992* (Franklin, W. Va.: The Mountain Institute, 1993); Task Force for the Makalu-Barun Conservation Project, *The Makalu-Barun National Park and Conservation Area Management Plan* (Kathmandu, Nepal, and Franklin, W. Va.: Department of National Parks and Wildlife Conservation, His Majesty's Government of Nepal, and The Mountain Institute, 1990); Bob Davis, senior program officer, The Mountain Institute, Franklin, W. Va., private communication, October 19, 1994; migratory corridors from Rodney Jackson, Wang Zhongyi, and Lu Xuedong, "Mountain Protected Areas and Snow Leopards: The Role of an Indicator Species in Reserve Design and Management," in *Proceedings of the First Conference of East Asian Protected Areas*, sponsored by CNPPA-IUCN and Chinese Academy of Sciences, September 12-16, 1993, Beijing, China, in press.

96. Reed F. Noss, "A Native Ecosystems Act," *Wild Earth*, Spring 1991; Noss and Cooperrider, op. cit. note 12.

97. Jack D. Ives, "The Mountain Malaise," in Singh and Kaur, op. cit. note 92; Jack D. Ives and Bruno Messerli, "Progress in Theoretical and Applied Mountain Research, 1973-1989, and Major Future Needs," *Mountain Research and Development*, May 1990; Jack D. Ives, "Institutional Frameworks for the Study of Mountain Environments and Development," *GeoJournal*, May 1992; Ives, "The Future," op. cit. note 9; Kelly A. McDonald and James H. Brown, "Using Montane Mammals to Model Extinctions Due to Global Change," *Conservation Biology*, September 1992; African Mountain Association (AMA), Mt. Kenya Ecological Programme (MKEP), Annex I, Third International AMA Workshop Proceedings, Nairobi, Kenya, 1993; Francis F. Ojany, president of African Mountain Association and professor of geography, University of Nairobi, Kenya, private communication, July 27, 1994; Andes from Li Pun, op. cit. note 18; Swiss program from Bruno Messerli, "Mountain Environments—A Reaction," *GeoJournal*, September 1994.

98. Stone, op. cit. note 1; Ives, op. cit. note 55.

99. CIPRA, op. cit. note 54; as of a ministerial meeting held in Chambéry, France on December 21, 1994, Austria, Liechtenstein, Germany, and the European Union had ratified the convention—a sufficient number to bring the treaty into legal force, from Andrea Dallinger, CIPRA, Munich, Germany, private communication, January 10, 1995.

100. ICIMOD, *Proceedings of the Tenth Anniversary Symposium of the International Centre for Integrated Mountain Development (ICIMOD)* (Kathmandu, Nepal: 1994); Egbert Pelinck, director general, ICIMOD, Kathmandu, Nepal, private communication at Worldwatch Institute, Washington, D.C., September 23, 1994; Jack D. Ives, Larry S. Hamilton, and Kevin O'Connor, "Feasibility Study for an International Mountain Research and Training Centre," *Mountain Research and Development*, May 1989.

101. Ecotrust, *Fiscal Year 1993 Annual Report* (Portland, Ore.: 1993); Jack and Hal Brill, "First Ecobank Connects Chicago to Pacific Forests," *In Business,* September/October 1994; Erin Kellogg, director, Policy and Communication, Ecotrust, Portland, Ore., private communication, November 2, 1994.

102. Figures for total World Bank lending from World Bank, *Annual Report* (Washington, D.C.: various years); another five mountain projects totalling $2.19 billion were for hydroelectric power projects, from Information Services Division, Operation Policy Department, World Bank, Washington, D.C., unpublished printout of internal project summaries, June 1994; records of other institutions from Jane Pratt, president and CEO, The Mountain Institute, Franklin, W. Va., private communications, September 27 and October 19, 1994; International Fund for Agricultural Development (IFAD), *1993 Annual Report* (Rome: 1994); Nessim Ahmad, Environment Adviser, IFAD, Rome, in Minutes of "Ad-Hoc Inter-Agency Meeting on UNCED Agenda 21, Chapter 13, *Managing Fragile Ecosystems: Sustainable Mountain Development,*" FAO, Rome, Italy, March 21-22, 1994.

103. John Clark, *Democratizing Development* (West Hartford, Conn.: Kumarian Press, 1991).

104. Mohammed T. El-Ashry, "The New Global Environment Facility," *Finance & Development,* June 1994; Russell A. Mittermeier and Ian A. Bowles, *The GEF and Biodiversity Conservation: Lessons to Date and Recommendations for Future Action,* Conservation International Policy Papers 1 (Washington, D.C.: Conservation International, May 1993); Frederik van Bolhuis, senior economist, Global Environment Facility, World Bank, Washington, D.C., private communication, September 28, 1994.

105. Thomas F. Homer-Dixon, Jeffrey H. Boutwell, and George W. Rathjens, "Environmental Change and Violent Conflict," *Scientific American,* February 1993; Robert Goodland, *Population, Technology, and Lifestyle* (Washington, D.C.: Island Press, 1992); Robert J.A. Goodland, Herman E. Daly, and Salah El Serafy, "The Urgent Need for Rapid Transition to Global Environmental Sustainability," *Environmental Conservation,* Winter 1993; Stephen C. Longeran, "Impoverishment, Population, and Environmental Degradation: The Case for Equity," *Environmental Conservation,* Winter 1993; Durning, op. cit. note 50.

PUBLICATION ORDER FORM

No. of
Copies

_____ 57. **Nuclear Power: The Market Test** by Christopher Flavin.

_____ 58. **Air Pollution, Acid Rain, and the Future of Forests** by Sandra Postel.

_____ 60. **Soil Erosion: Quiet Crisis in the World Economy** by Lester R. Brown and Edward C. Wolf.

_____ 61. **Electricity's Future: The Shift to Efficiency and Small-Scale Power** by Christopher Flavin.

_____ 63. **Energy Productivity: Key to Environmental Protection and Economic Progress** by William U. Chandler.

_____ 65. **Reversing Africa's Decline** by Lester R. Brown and Edward C. Wolf.

_____ 66. **World Oil: Coping With the Dangers of Success** by Christopher Flavin.

_____ 67. **Conserving Water: The Untapped Alternative** by Sandra Postel.

_____ 68. **Banishing Tobacco** by William U. Chandler.

_____ 69. **Decommissioning: Nuclear Power's Missing Link** by Cynthia Pollock.

_____ 70. **Electricity For A Developing World: New Directions** by Christopher Flavin.

_____ 71. **Altering the Earth's Chemistry: Assessing the Risks** by Sandra Postel.

_____ 75. **Reassessing Nuclear Power: The Fallout From Chernobyl** by Christopher Flavin.

_____ 77. **The Future of Urbanization: Facing the Ecological and Economic Constraints** by Lester R. Brown and Jodi L. Jacobson.

_____ 78. **On the Brink of Extinction: Conserving The Diversity of Life** by Edward C. Wolf.

_____ 79. **Defusing the Toxics Threat: Controlling Pesticides and Industrial Waste** by Sandra Postel.

_____ 80. **Planning the Global Family** by Jodi L. Jacobson.

_____ 81. **Renewable Energy: Today's Contribution, Tomorrow's Promise** by Cynthia Pollock Shea.

_____ 82. **Building on Success: The Age of Energy Efficiency** by Christopher Flavin and Alan B. Durning.

_____ 83. **Reforesting the Earth** by Sandra Postel and Lori Heise.

_____ 84. **Rethinking the Role of the Automobile** by Michael Renner.

_____ 86. **Environmental Refugees: A Yardstick of Habitability** by Jodi L. Jacobson.

_____ 88. **Action at the Grassroots: Fighting Poverty and Environmental Decline** by Alan B. Durning.

_____ 89. **National Security: The Economic and Environmental Dimensions** by Michael Renner.

_____ 90. **The Bicycle: Vehicle for a Small Planet** by Marcia D. Lowe.

_____ 91. **Slowing Global Warming: A Worldwide Strategy** by Christopher Flavin

_____ 92. **Poverty and the Environment: Reversing the Downward Spiral** by Alan B. Durning.

_____ 93. **Water for Agriculture: Facing the Limits** by Sandra Postel.

_____ 94. **Clearing the Air: A Global Agenda** by Hilary F. French.

_____ 95. **Apartheid's Environmental Toll** by Alan B. Durning.

_____ 96. **Swords Into Plowshares: Converting to a Peace Economy** by Michael Renner.

_____ 97. **The Global Politics of Abortion** by Jodi L. Jacobson.

_____ 98. **Alternatives to the Automobile: Transport for Livable Cities** by Marcia D. Lowe.

_____ 99. **Green Revolutions: Environmental Reconstruction in Eastern Europe and the Soviet Union** by Hilary F. French.

_____ 100. **Beyond the Petroleum Age: Designing a Solar Economy** by Christopher Flavin and Nicholas Lenssen.

_____ 101. **Discarding the Throwaway Society** by John E. Young.

_____ 102. **Women's Reproductive Health: The Silent Emergency** by Jodi L. Jacobson.

_____ 103. **Taking Stock: Animal Farming and the Environment** by Alan B. Durning and Holly B. Brough.

_____ 104. **Jobs in a Sustainable Economy** by Michael Renner.

_____ 105. **Shaping Cities: The Environmental and Human Dimensions** by Marcia D. Lowe.

_____ 106. **Nuclear Waste: The Problem That Won't Go Away** by Nicholas Lenssen.

_____ **Total Copies**

☐ **Single Copy: $5.00**

☐ **Bulk Copies (any combination of titles)**

   ☐ 2–5: $4.00 ea.   ☐ 6–20: $3.00 ea.   ☐ 21 or more: $2.00 ea.

   Inquire for discounts on larger orders.

☐ **Membership in the Worldwatch Library: $30.00 (international airmail $45.00)**

The paperback edition of our 250-page "annual physical of the planet,"
State of the World, plus all Worldwatch Papers released during the calendar year.

☐ **Subscription to *World Watch* Magazine: $20.00 (international airmail $35.00)**

☐ **Worldwatch Database Disk: $89**

Includes up-to-the-minute global agricultural, energy, economic, environmental, social,
and military indicators from all current Worldwatch publications.

Please check one: _____high-density IBM-compatible or   _____Macintosh

---

Stay abreast of global environmental trends and issues with our award-
winning, eminently readable bimonthly magazine.

Please include $3 postage and handling for non-subscription orders.

Make check payable to Worldwatch Institute

1776 Massachusetts Avenue, N.W., Washington, D.C. 20036-1904 USA

Enclosed is my check for U.S. $_____

VISA ☐   MasterCard ☐ _____

                           Card Number   Expiration Date

---

name   **daytime phone #**

---

address

---

city   **state**   **zip/country**   WWP

Phone: (202) 452-1999   **Fax: (202) 296-7365**   **E-Mail: wwpub@igc.apc.org**